A moored 'Lincoln' size vessel with an attractive display of carving on the hawse timber.

A Pictorial History of Canal Craft

Peter L Smith

B T Batsford Ltd London

First published 1979

ISBN 0 7134 1637 8

Filmset in 'Monophoto' Photina by
Servis Filmsetting Limited, Manchester

Printed by The Anchor Press Limited, Tiptree, Essex
for the publishers B T Batsford Limited
4 Fitzhardinge Street, London W1H 0AH

Contents

Acknowledgment

The author is grateful to the following people and organisations for their kind permission to reproduce the listed photographs: Stewart Bale, Liverpool for 34; Barge & Canal Development Association for 2, 5, 7, 16, 41, 54; Bridgewater Canal Company for 32, 37, 38, 46, 57, 59; Bude-Stratton Town Council for 49; City Library, Worcester for 61, 62, 63; County Library, Gainsborough for 1, 4, 6, 87; County Library, Nottingham for 10, 108; County Library, Reading for 71; *Daily Mail* for 80; District Central Library, Accrington for 100; H Draycott, The Fowler Collection/Museum of English Rural Life, University of Reading, for the endpaper photograph; Dudley Libraries for 93, 94; Richard Dunston Ltd for 22; E A Edwards for 8, 109; Alan Faulkner Collection/John W Bell for 77; Alan Faulkner Collection/Cambridge & County Folk Museum for 78; Alan Faulkner Collection/Eastern Gas Board for 79; Alan Faulkner Collection/A Leach for 40; Alan Faulkner Collection/Miss E Murrell for 76, 110; Gladwin & White for 91, 105; John Harker Ltd for 26; Ironbridge Gorge Museum Trust for 48; Lancaster Museum for 30, 106; Leeds City Libraries for 11; Linlithgow Union Canal Society/Falkirk Library for 67, 68, 69, 107; Liverpool City Libraries for 53, 55; Robert May for 84, 97, 99; Peter Norton for 56, 58, 101; British Waterways Board for 27; P J G Ransom for 65; Records Office, Hertford for 36, 86–8; Rochdale Libraries for 82; Sheffield City Libraries for 9; J G Simpson for 17, 20, 25, 29; W Walker for 15; Waterways Museum, Stoke Bruerne for 74, 83; Alan West for 33; West Riding Boat Co Ltd for 117; John H Whitaker Ltd for 13, 21; D Wood for 73, 75; D Wood/Richmond Public Library for 70; Elizabeth Wood for 72; *Yorkshire Post* for 31, 39, 50.

The following photographs are from the author's collection: frontispiece (p. 6), 3, 12, 42, 45, 64, 104, 112, 113. The following photographs are the work of the author: 14, 18, 19, 23, 24, 28, 35, 43, 44, 47, 51, 52, 60, 66, 81, 85, 89, 90, 92, 95, 96, 98, 102, 103, 111, 114, 115, 116, 118.

The author is also grateful to John W Holroyd for the drawings and artwork, and Andrew Rhodes for the copy drawing of the compartment steam tug on pages 40–1. Valuable assistance has been received from: A Littlewood, the Manager of the Bridgewater Canal; A J Conder the Manager/Curator of the Waterways Museum; Edward Paget-Tomlinson, David Wood, R J K Murphy, D D Gladwin of the Waterways Research Centre; S B Smith of the Ironbridge Gorge Museum Trust; Miss L Cooling, Alan H Faulkner, John Goodchild, Peter Norton, Geoff Wheat of Northern Counties Carriers Ltd; Eric Grubb of John H Whitaker (Tankers) Ltd; Miss S Doeg of *Waterways News*; and finally Barge & Canal Development Association members Adrian M Simpson and Jeremy G Simpson, who provided much enthusiastic and copious assistance.

Introduction

THE FACT that this book is entitled *A Pictorial History of Canal Craft* may incline the reader to presuppose that it is confined to describing and illustrating only the vessels that for nearly two centuries worked on the comparatively still waters of canals alone. But the book does cover a wider sphere, including vessels that did, and still do, work onto other inland waterways, the rivers and navigations. Without these craft the book would not be complete, for really very few types of craft were confined to canals alone, the majority working as a matter of course onto rivers and navigations. In fact some types of canal vessels were so closely allied to craft that worked on so-called 'moving waters' that it is difficult to differentiate between them, while many have their beginnings, and owe their development to operations on waterways other than pure canals.

The inland waterways were the lines of communication that constituted the backbone of Britain's industrial revolution on which much of the nation's prosperity relied for many years. The history of the craft extends from the beginning of commercial transportation on inland waterways through to the present day; commercial transportation is now basically confined to the network of waterways extending inland from the main estuaries, and in particular the Humber estuary.

This book does not specialise on sea-going coastal vessels, although many of the small sailing barges that in the past did venture onto coastal waters are included. Most of the boats and barges were built on sea-going principles, with a keel, and frames to shape the planking that ran longitudinally from the bilges upwards. From one period to the next, the vessels experienced the same changes in construction, from wood to iron and steel, and the abandonment of riveting in favour of welding.

What is the deciding factor relating to the classification of a vessel's grouping – when is it a boat and when a barge? The simple answer is width; a barge is a canal or river vessel not less than 11 feet wide, anything narrower is a boat. One exception of note may be the vessels of the Leeds & Liverpool Canal that have always been referred to as boats despite the fact that they are of barge dimensions.

In addition to the many cargo carrying vessels, the book includes details of maintenance craft, passenger boats, and, because of the vast increase in use of canals by the public in recent years, pleasure craft. Many people have their own favourite group of vessels, and study or are intrigued by some particular type of craft. Indeed there are so many variations in detail and design of certain groups of vessels it is easy to over-simplify by grouping, and if I have not included sufficient material on the reader's particular favourites, then I apologise.

Never before has so much interest been shown for our canal system and inland waterways in general. To cater for this thirst for knowledge about them, a continual stream of books has been published in recent years. It is now becoming more and more difficult to locate old prints that have not been published previously, and although I have endeavoured to include as much unpublished material as possible, I have had to include several that have been published before to illustrate certain specific aspects; it is hoped that the reader will accept these few repeats as they are intended, because they are of irretrievable historical value.

1 This 'Lincoln' size, sloop-rigged vessel on the River Trent has a favourable wind, so the leeboard is not in use, yet it can be clearly seen in the rest position, secured alongside the hull. In tow is the coggy boat, the small rowing boat that was part of the equipment of virtually all sailing vessels.

Vessels of the North East

THE HUMBER estuary, with its adjoining tidal rivers, was for many hundreds of years the operating area of small sailing vessels. Originally these transported agricultural products and raw materials such as sand and stone to the growing towns and cities in the area. Over the years the vessels, which had to cope with wind, tide and sandbanks, gradually developed as flat-bottomed craft, with round bows and sterns. The flat bottom was useful on tidal rivers, for when the vessels were stuck on a sandbank, or discharging at a riverside wharf as the tide ebbed, they were able to stay upright without any fear of turning over. In the late 1600s these vessels already had established trading on the rivers Humber, Ouse, Trent and around the coastal region, and also as far along other rivers, like the Wharfe, Aire, Hull, Idle and Don, as they could go before their progress was impeded by the many mill weirs and shallows.

The pound lock, first used in Britain on the Exeter Canal in the 1560s, provided a satisfactory method of by-passing weirs, thus enabling river navigations to be developed. The first river improvement of note in the area was the Aire & Calder Navigation, which made the rivers navigable for the first time to Leeds in 1700 and Wakefield in 1701. The same vessels that worked on the tidal waters naturally worked on the navigations and canals, which continued to expand inland throughout the next century.

On the tidal waters the Yorkshire keels had relied upon a large square sail, but because of the bridges and locks on the inland waterway, their appearance inland was not as common, although they were used in favourable conditions. Men were employed for a time to bow haul them, as they had done on the rivers, when the wind was not favourable. The use of the horse had generally

Examples of rigging of North Eastern vessels (not to scale with each other)

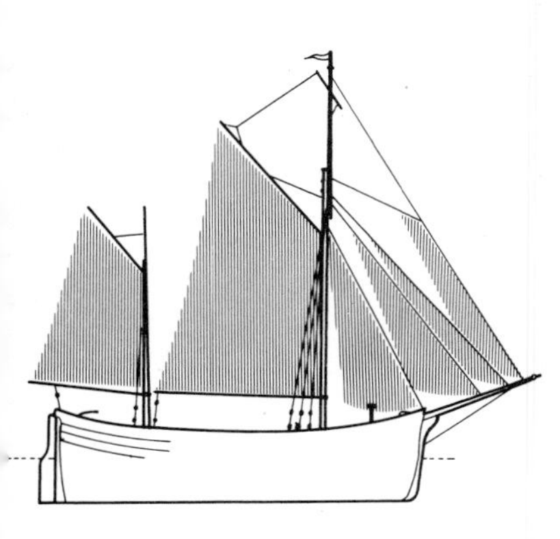

Billy Boy rigging

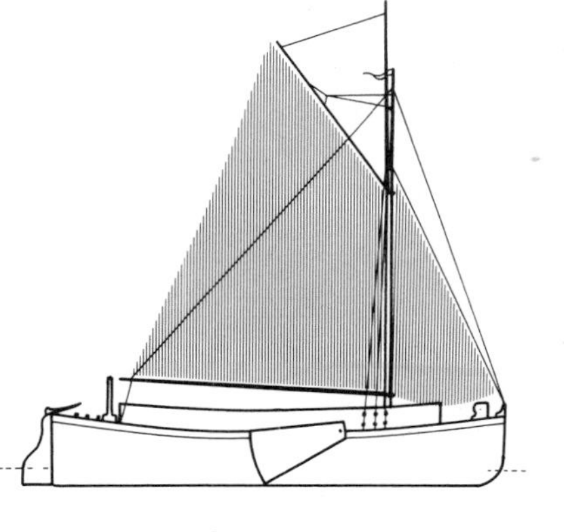

Sloop rigging

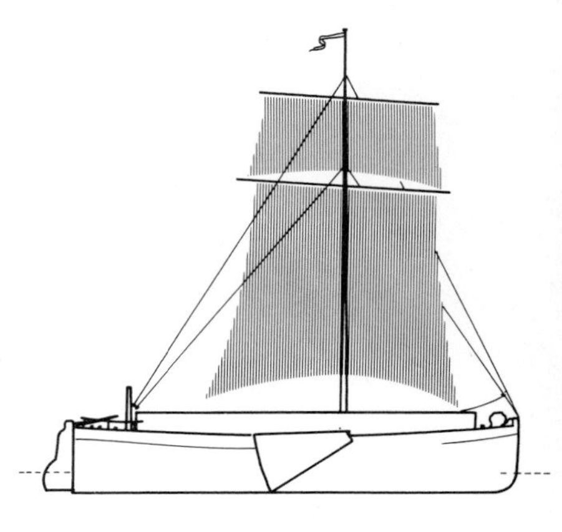

Keel rigging

2 This wooden Yorkshire keel under sail at Hull Bridge near Beverley, in 1935, was en-route to Driffield. With lines ashore, the vessel was waiting for fresh water from the flooding tide. Above the stem on the hawse timber can be seen a carved motif, centre of the two eyes, so typical of the conservative decoration of these vessels.

become widespread to haul the keels, but from the mid-nineteenth century onwards, the horse was replaced in part by the steam tug, which was able to haul several keels at a time.

At the time of the expansion of the inland waterways the majority of the keels were about 55 feet in length with a beam in the region of 14 feet, although some were a few feet longer and a few inches wider. The locks on the early navigations were made to accommodate the size of vessel common in the immediate area. Keel hulls of clinker construction, with planks overlapping, were common at first, but over the years several changes took place until, a little before the turn of the last century, carvel construction (that is, smooth-sided, with one plank butting up against its neighbour) became common building practice. The building yards were usually small family concerns, building on average only one or two new vessels each year, and relying mainly on repair work. There were dozens of such yards throughout the network of waterways in the area.

3 This unloaded sloop at Tetney lock, the entrance of the Louth Canal, is of clinker construction, a method which had already lost popularity in favour of carvel construction by the time this photograph was taken. Note the full rounded stern; the timbers for this required much steaming, along with the full expertise of the builder to construct such a shape from planks of oak 3 or more inches thick.

The Yorkshire keels had well rounded, rather bluff bows, with sterns a little less rounded than the bows to allow the flow of water to the rudder. The hulls were tarred to a little above water line, to give the best protection to these strongly made vessels. The planks of oak were some $2\frac{1}{2}$ to 3 inches thick, placed on ribs that were up to about 8 inches thick. This gave them great strength, for no matter how skilfully the vessels were handled, occasional knocks and scrapes could not be avoided, especially when working into locks from tidal waters. This was extremely difficult for vessels without any mechanical power to assist or arrest their movement, and in fact at some tidal locks winches were provided to haul the vessels into the lock. Around the hull at both the bow and

4 Sailing keels with their masts lowered are hauled up the River Trent by a steam-powered tug. Similar workings with large dumb craft and diesel tugs continued until the mid 1970s from Hull to a flour mill at Gainsborough.

stern were fitted rubbing bands, strips of metal on a foundation of oak, to provide a raised surface to take the wear when the vessel rubbed against an obstruction.

The main structures of the vessel were the keel, keelson, ribs, front stem and rudder post. The timber for the stem was on many occasions selected by the boatbuilder as a curved bough still growing on a tree. This ensured that the run of the grain retained the strength needed for such an important fixture. The bottom of the vessel was usually of elm, for the properties of this wood, while continually submerged in water, are excellent. These basic building practices were not confined to keel construction, but used at yards throughout the country.

The main consideration was the space for the cargo in the hold. This was large and spacious and had supporting cross members, at deck level. An inner lining was fitted into the hold, and at intervals around the sides there were inspection panels to enable the interior of the vessel's hull to be inspected. The floor of the hold could be removed on occasions to allow for the removal of the residue from bulk cargoes that had penetrated into the bilge space below the floor. It was essential to keep the bilges clean, to allow water to run to pump suction points, and to allow air to circulate. To enable the cargo hold to carry as much as possible, it was extended upwards, perhaps as much as several feet, by means of fixed planking above deck level. At the sides these planks were known as coamings and at the front and back as headledges. The hold was covered by loose panels of boards that could easily be removed for loading or discharging. Then it was covered overall by tarpaulin sheets, which were roped down to keep the weather out.

5 Repairs and renewals: Stanisland's yard at Thorne on the Stainforth & Keadby Canal, with wooden vessels under repair, showing clearly the ribs of the vessels' frame. Some of the new ribs may probably look rather too long, especially on the vessel on the left. When the job is finished they will be new timber heads for the securing of mooring ropes. In the foreground is a discarded piece of rubbing band as well as a keel mast.

The deck of the vessel, both fore and aft, was of scrubbed white planking, with access to either end gained by means of a catwalk down each side of the coamings of the hold. The catwalks or side-decks ranged in width from a few inches to a couple of feet. Two living cabins, one fore and one aft below decks, provided accommodation for all the family, for the boats were certainly family boats. Father was skipper, mother usually assisted as first mate if a man was not employed, and the children assisted in the running of the vessel from an early age. As the craft was on the move most of the time, very few of the children attended school, and so many were unable to read or write. The skippers did not use banks but relied upon their own back trouser pocket, paying the tolls and for services and supplies as they went.

At the front of the vessel the stem protruded forward from the hull, perhaps by as much as 2 feet, and this was protected by several rope fenders. A large one, a bow fender, was at the centre, and there would be at least one smaller one, some distance below, known as a cill fender, protecting the stem from direct contact with the cills of locks. Several hanging fenders, similar to but thicker than bell rope ends, hung over the bow and stern, to give further protection to the hull. The fenders, of plaited rope, were usually made by the skipper or one of the older children.

At the bows, extending upwards several feet above the line of the hull and deck, was the hawse timber. This was of solid construction and was usually adorned with ornate carvings, worked by the craftsmen builders, to give a little touch of distinction. These included a central motif, which varied considerably from rising suns and beehives to crossed swords, etc. On some vessels an impressive motif was also to be found on the rudder stock. The carvings usually

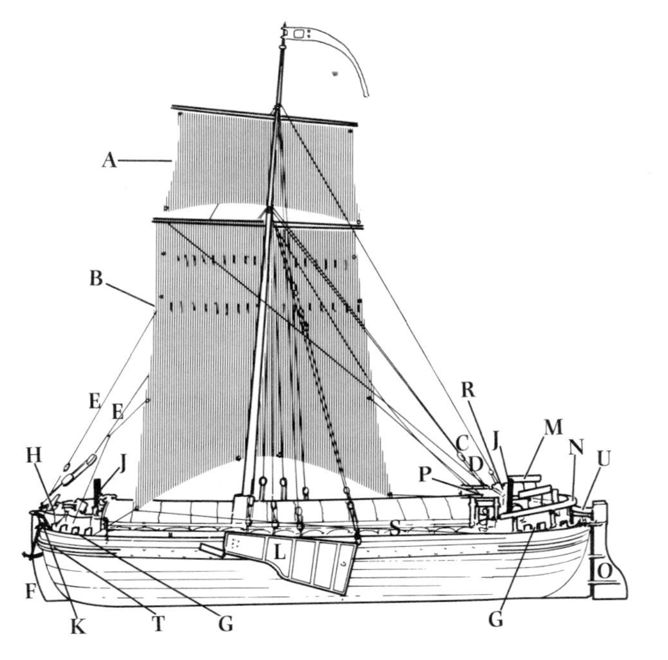

Yorkshire keel, 'Sheffield' size, 61ft 6in long

Key:
A Top sail
B Main sail
C Gin block
D Slabline halyards
E Forestays
F Stem
G Timber heads
H Winch
J Stove chimney
K Hawse timber
L Leeboard
M Navigation light boards
N Stern rail
O Rudder
P Sheet rollers
R Tack rollers
S Coamings
T Top strake
U Tiller

had gold leaf applied to enhance their appearance even further.

Behind the hawse timber, located on the forward deck, was a large winch with an exceptionally strong wooden barrel with metal bands. This was operated by a handspike and was provided to lift the anchor when it had been lowered on a tidal section. Alongside the hawse timber, following the line of the deck, a number of stanchions were located. These were known as timber heads, to which the mooring ropes were secured.

A strong rail of oak, steamed and carved by the builders, surrounded the stern deck, and on it was a little delicate carving, to match the hawse timber at the bows. The carving usually included the boat's name, owner and registration details. On the stern deck, in addition to a hatch cover that gave access to the stern living cabin, was a barrel for the storage of drinking water.

The mast, positioned a little forward of midships, was secured in its base unit, known as the lutchet or tabernacle. It could be lifted or lowered as required to enable the vessel to pass under low bridges. To complete the rigging equipment were the ropes, sails, rollers, dead-eyes, blocks, and winches. On either side of the hull were the leeboards, made of wood and similar in shape to butterfly wings. Secured at the top, they could be lowered into the water to act as stabilisers against the sideward movement of the boat, especially when the wind blew from the side. To a certain extent this prevented the craft being blown off course, and so made it easier for the helmsman to steer accurately.

The cabins, with the steps leading down from the deck hatch, had three sides of the floor space fitted with cupboards from floor to ceiling. The side cupboards included two that gave access to bed spaces that were complete with mattresses, pillows and blankets. They were usually adequate with enough headroom for an adult to sit up in. Above each was a deck light, a small glass panel in the deck that allowed just sufficient daylight to see by. During daylight

6 Two Yorkshire keels at Keadby, the junction of the Stainforth & Keadby Canal of the Sheffield & South Yorkshire Navigation with the River Trent. The one with the sail hoisted is of carvel construction; these are family boats, and the husband and wife on the aft deck are ready to cast off. *Brasso* on the right is of steel construction, and the sails are stored in their special cover. This vessel was later converted to a motor barge, and is still in existence complete with wheelhouse. On the left, underway, is *Jim*, an early John Harker 'Sheffield' size vessel, tiller-steered and of steel construction, used for transporting bulk fuel cargoes.

hours the light gained access to the main cabin area from two further small deck lights in addition to the open hatchway, while at night the cabins were illuminated by a hanging copper paraffin lamp.

The cupboards on the third and intermediate section, towards the actual stem or rudder post, provided storage space for stores and provisions, and in front of this section there was usually a folding table. Around the base of the cupboards were additional lockers and storage space, fitted with bench tops for seating. Frosted glass panels in the numerous cupboard doors helped to use the limited light to the best advantage. The fourth side of the living cabin, the side nearest the cargo hold, was usually where the stove was positioned, sitting on a slab of stone. The stove had a little shelf called a 'Tidy Betty' in front of the oven. It burned coal and was used for cooking and heating purposes. Made of cast iron, with inlaid patterns of leaves and scrolls, it was always kept well black leaded, with a polished copper front ash pan below.

The cabin floor was covered by either a clipped rug or washed sacks. A few personal pictures adorned the cabin wall near the stove, but the ceiling was painted white, in contrast with the fitted cupboards, which were stained and varnished and sometimes decorated with carvings. The cabins were warm and very cosy to live in, with fresh air entering from one or both of the ventilators and if need be from the opened hatchway above the access steps. When the

7 *Hannah* was a 'west country' size Yorkshire keel. The intricate carving of the stern rail and rudder can be clearly seen. The metal tiller is for canal work, while the large sweeping wooden tiller was usually used on rivers.

vessel was underway, especially on tidal waters in bad weather, say from Hull heading inland, the hatch usually had to be closed. Otherwise, water from the waves lashing over the deck would pour down into the cabin. On occasions the vessels would have to moor alongside jetties on tidal waters. At night at such times when the weather was bad, it could be difficult for the crew to sleep as they were continually disturbed by numerous creaks and groans from the vessel and the rattle of the paraffin lamp.

It was a hard life in many ways, with all the elements to contend with, especially in the winter time, in temperatures below freezing. Then it was a struggle with wet and frozen sails and ropes, standing for hours in the biting wind steering the vessel with the 9 foot long tiller. On the canal sections, at times when they were unable to pay for a hired horse or tug, the family crew had to walk on the towpath and bow haul the vessel, perhaps for a whole day or more. From an early age the boys learned about the working of the vessel, and about the rivers' different moods, shallows and channels. Despite all the difficulties, however, the keelmen had the satisfaction of a job well done, and this brought contentment for them and their families. Of course life had its lighter side, with annual regattas and galas, keel racing and rowing competitions. These were not confined to the menfolk alone, as the women competed in many events. These competitions were very popular, with many cups and challenge shields being presented to the winners.

The skipper and his family were proud of their vessel, be it their own or a company boat, and always ensured that it was turned out in the best of trim, with all ropes neat and tidy. The paintwork was bright and clean, for the coamings, headledges, timber heads, stern rail, hawse timber and cabin hatches were always painted in bright colours. Some even had certain sections grained and varnished instead of painted, and these would include the top plank of the hull, especially at the fore and aft sections, known as the top strake.

A Yorkshire keel was a vessel with a large square-rigged sail, with, on occasions, a smaller top sail. In addition there were vessels with exactly the same hull, but sloop rigged; these were Yorkshire sloops, once in general use throughout the area – in their latter days of operation they were favoured by owners who resided on the Lincolnshire coast of the Humber estuary and on the waterways within that county.

Also in operation in the North East for many years were the Billy Boys. These were capable of carrying 80 to 100 tons, built on similar lines to the keels, but they were deeper draughted, all of clinker construction, with more shear, and bluffer bows. Like the keels and sloops, they were fitted with leeboards, but as they were about 63 feet long with a beam of about 18 feet, they were too large to work off the tidal rivers until the Aire & Calder Navigation enlarged its locks. Then they were occasional visitors to both Leeds and Wakefield. Basically sea-going vessels, being too large for the majority of inland waterways, they continued in operation until the turn of the last century when their numbers decreased considerably.

After the opening of the navigations and canals, the sailing vessels of the North East worked north as far as Ripon. They went to Tadcaster on the River Wharfe, to Driffield on the Driffield Navigation, as well as to Beverley on the Beverley Beck, and the River Foss from York. They also worked to Leeds and

8 A Billy Boy, of clinker construction, is unloaded at Spalding on the River Welland in 1900, and is no doubt ready to return to the River Humber.

9 A 'Sheffield' size Yorkshire keel, complete with top sail, successfully negotiates the narrows of the swing railway bridge on the Stainforth & Keadby Canal.

Wakefield on the Aire & Calder Navigation, and Sheffield on the River Don Navigation that was later to become the Sheffield & South Yorkshire Navigation and Bawtry on the River Idle.

They were also the most common craft on other Yorkshire waterways such as the Barnsley, Selby, Market Weighton, and Dearne & Dove Canals. In addition to the Calder & Hebble Navigation, from where they travelled along the Rochdale Canal into Lancashire and, via Runcorn, to Liverpool, a few used the Leeds & Liverpool Canal route to Liverpool. These vessels also worked to Lincoln on the Fossdyke, and beyond along the River Witham as well as on the Sleaford Navigation. They penetrated up the River Trent to Nottingham and a little beyond.

Trent craft before 1805 worked as far as Burton, and thereafter to Wilden Ferry. There were two distinct sizes: the Upper Trent craft that went above Nottingham, and kept below Keadby, and the Lower Trent craft that worked between the Humber and Nottingham – these latter were keels in the accepted sense. After the improvement of the Trent in the early 1800s a minimum depth of 2 feet 6 inches was maintained for many years. Above Nottingham the locks determined a maximum size of 72 feet by 14 feet, but vessels based on the keels, usually 60 feet by 11 to 14 foot beam, operated carrying a maximum of 50 to 60 tons, working up the River Soar or on the Nottingham canals until the late 1950s when traffic terminated above Nottingham. Over the years a number of locks on the River Trent below Nottingham were enlarged to 180 feet by 29 feet, but as a number of small locks 81 feet in length remained until 1954, it was not until after that period the large modern vessels 142 feet by 17 feet 6 inches, carrying up to 250 tons, could reach Nottingham.

The wooden-constructed keels totalled many thousands at any one time, and had one of the largest operating areas in the country. There were many variations in size, the smallest of them having a cargo capacity of between 40 and 50 tons. The largest were the Humber keels that were usually a foot deeper in the hold than the Yorkshire keels. The Yorkshire keels were built to the maximum size that a waterway's locks could accommodate, and were known usually by that localised size, e.g. Lincoln, Sheffield, Beverley, Barnsley size, etc. Until the middle of the last century, the 57 foot 6 inch long keels, for which the locks of the Calder & Hebble Navigation were constructed, were the most common size, and also one of the smallest size in operation. Although over the years many waterways enlarged their locks, those on the Calder & Hebble on the whole were never enlarged, so the original small-sized keels continued to work on this waterway. They became known as 'west country' vessels, that is vessels that worked to the west of the wide waterways of the Humber area.

The basic hull design of the keel was adapted and utilised for much localised work, as were the Upper Trent barges. In addition, other vessels were constructed for 'internal' work on certain waterways, usually transporting one type of commodity only, and in some cases on day work only, manned by men without their families. One such type used for transporting coal became known as a coal slacker. Another type was the Mallion boat, used for transporting heavy stone material on the Aire & Calder Navigation. The majority of these 'specials' were built without any intention of their doing tidal work or being required to provide accommodation, and were designed on

10 These two types of Trent barges alongside Trent Navigation warehouse at Nottingham in 1929 include the shallow-draughted type used for working further inland, up as far as the River Soar.

11 This scene shows the River Aire in the centre of Leeds. Alongside the Aire & Calder warehouse on the right is moored a steel-hulled 'Sheffield' size sloop *Clarence T*, built by Warrens Shipyard. The vessel was later converted to a motor barge with wheel steering. The vessel, always owned and operated by the Tomlinson family, continued to carry cargoes until June 1972 when it was cut up for scrap. Its last regular working was grain from Hull to Sheffield. Also in the photograph can be seen a tug, two dumb barges, and, alongside the wharf on the left, a large capacity steam barge or packet – only a few were built for operation in the north east.

12 In later years, the Aire & Calder Navigation Company owned nearly one hundred of these general cargo vessels, which were mostly hauled by steam tugs. The photograph shows the rescue of craft from an unnavigable section of the River Calder; the two vessels had been washed over the Kirkthorpe weir by flood water, and taken several miles along the river to the breakwater of the Stanley Ferry Aqueduct. The vessels, resting only a few yards from the navigation, were dragged up the river bank and launched into the canal section once more. Because of their strong construction they were able to return to work immediately despite the undue stress and strain to which they had been subjected.

13 Accidents do happen; here two large capacity dumb vessels, one of steel and the other of wood, have come to grief in a lock at Hull.

14 A wooden, 'West Country' size, keel-hulled motor barge, built at Mirfield on the Calder & Hebble Navigation in 1952, *Ethel*, is the last wooden barge in Yorkshire to have carried commercially. Originally built as tiller-steered for the Calder Carrying Company, later, when in the fleet of Hargreaves (West Riding) Ltd, she was converted to wheel steering. Withdrawn from service early in 1975, she is now the only example of a Yorkshire vessel that the Boat Museum at Ellesmere Port have.

15 Definitely a sad way to go, but at least a final one, without being an obstruction to navigation.

16 *Ethel*, dressed overall with flags and bunting at her launching day on 28 July 1952 at Mirfield on the Calder & Hebble Navigation.

17 A steam packet at the riverside at Wakefield in the early 1930s: although not many of these steel-constructed vessels were built, one converted to diesel power remained at work until the early 1970s.

simple lines, with the largest capacity hold possible, some even having a transom stern similar to Leeds & Liverpool vessels.

During the latter part of the nineteenth century the first vessels constructed of iron made their appearance on the waterways of the North East. The new metal vessels retained the basic shape and fittings of the wooden craft, losing only a little of the carved wood decoration. From then on iron, and later steel, became a regular building material, and gradually became a common sight. Some operators remained loyal to timber-constructed vessels, and although increasingly over the years more and more metal barges were built, it was not until 1955 that the last timber-constructed vessel was launched, still with the original keel hull.

From the 1830s steam propulsion units were increasingly installed in vessels; a few of the former sailing keels were converted to steam, and a number of cargo carrying craft were built especially to accommodate the steam power unit. Because the use of tugs became common practice on many of the local waterways, the steam keels were not a common sight. The Aire & Calder Navigation, however, built several for special workings – in fact one remained in use operating between Goole and Barnsley until 1950. When the internal combustion engine took the place of steam, these engines were increasingly installed in former sailing keels as well as new vessels. Then, when the diesel engine proved to be more reliable in operation, it became standard practice to instal these.

Despite this, the small dumb vessels continued to be used, only finally ceasing with the closure of the Leeds Cooperative waterborne coal activities in the early 1970s. This company was the last inland waterway user in the area of a fleet of dumb vessels; a couple of operators who concentrate their activities on the tidal rivers in the area still use them today, although they are large 200 ton capacity

18 A motor barge is underway on the Calder & Hebble Navigation, the last internal working on the waterway using the smallest commercial barges in the north east. *Fidelity*, carrying 75 tons of coal, is made of steel; originally owned by J Wilby of York, it was built as a dumb craft. It was purchased by the present operator to replace worn out wooden, 'west country' size, keel-hulled motor vessels in 1971. Practically rebuilt and converted to a motor vessel, it entered service on the Navigation in October of the same year.

craft, and more like lighters than the original dumb vessels.

Metal motor barges, still based on the traditional lines and sizes, were increasingly constructed during the 1930s, along with some capable of carrying 150 to 200 tons for work on a few of the navigations in the area. These large vessels, including an increasing number of bulk fuel vessels, had a refinement unheard of before: a wheelhouse to protect the crew from the weather, and a wheel to steer with instead of the conventional tiller.

The first metal vessels were of riveted construction; this was a slow method of building, since each rivet had to be fitted individually by a squad of three men: the rivet heater, riveter and holder up. The first man heated the rivet, the second held the red-hot rivet in place with tongs, and the third man skilfully employed a hammer. In the early 1940s, the process was speeded up by the use of hand-held pneumatic hammers.

In the early 1940s the welding of vessels became standard practice. This speeded up the building, for various processes such as countersinking, riveting and joggling (bending the edges of metal) were dispensed with. In addition, the newly introduced gas cutting method replaced the previous lengthy process of shearing. The new methods of production resulted in the all-welded vessels

19 Part of a large fleet, these vessels in daily use on the Aire & Calder Navigation transport 220 tons of coal each. The fleet, all built in the late 1950s and early 60s, were first hauled by steam tugs, then during the mid 60s they were converted to motor barges.

20 The traditional lines of the 'Sheffield' size vessels changed in the 1940s with the introduction of modern building techniques. Here *Elspeth Mabel*, built in the 1950s, was one of the last of this special size to be built. She was of welded steel construction, incorporating a cut back, swim stern instead of the full rounded shape of the older vessels, as well as a full wrap round stern guard instead of rails. The vessel, owned by E V Waddington Ltd, is still in daily operation.

21 Accidents fortunately are not common occurrences, but a slight collision can result in expensive repairs. The torn hull and side decks of this tanker were the result of an impact from another vessel's bows.

22 *Hanley's Pride*, built 1937 by Richard Dunston Ltd of Thorne, is seen on test runs on the Humber before being handed over to her owners. Of riveted construction, and modern wrap round stern instead of rails, the cargo hold was a little smaller than normal, capable of holding 80 tons. This was to allow for an enlarged engine room to house a powerful engine, for the vessel was designed to haul several dumb vessels. The name was very appropriate, for she was pride of the company's fleet for many years. In October 1970 she was cut up for scrap, and her bottom was found to be nearly as thick as on the day she was built.

23 Bulk fuel now amounts to a substantial traffic on the north-eastern inland waterways, transported in large capacity modern vessels. In the first instance liquids were transported in barrels, then vessels of conventional design were fitted with lead tanks, followed by small specially built steel tank craft. Finally came the modern oil tanker, the largest with over 600 ton carrying capacity in use on the Aire & Calder Navigation. *Rufus Stone*, formerly owned by Cory Tank Craft Ltd, and built in 1963, is now operated by John H Whitaker (Tankers) Ltd.

24 In recent years, a boost to traffic figures has been achieved with the transport of effluent by barge. Initially, ex north-eastern oil tankers – such as *Sandall*, the empty craft – were used, but an increase in tonnage required an increased fleet, and as no further British vessels were available (except new ones) a number of vessels were brought over from Europe. Former Belgian and Dutch oil tankers such as the loaded vessels, *Howden*, *Esholt* and *Millshawe* (renamed after Yorkshire sewage works), are now regular users of the Aire & Calder Navigation.

25 *Dunlin C*, loaded, passes through Castleford Flood Lock on her way to Leeds. The vessel, built by Campling of Goole, has a gross registered tonnage of 174.55 tons.

26 *Farndale H*, 500 ton capacity, and 180ft long, is seen here being launched in 1967 at Knottingley by her builders and owners John Harker Ltd. Her first load was to Leeds from the Humber estuary in June 1967. In 1976, the vessel was sold to a new operator, Whitfleet Ltd, who continue to use her for deliveries to their bulk fuel depot on the Leeds section of the Aire & Calder Navigation.

27 This scene shows various sizes and types of vessels in use on the north eastern waterways in the mid 1960s. On the stocks, a new large capacity, inland waterway tanker is under construction. Alongside on a floating pontoon, one of the last wooden 'Humber' size keel-hulled vessels is under repair, while passing on the River Hull is a riveted constructed motor barge.

28 *Risby*, owned and operated by Flixborough Shipping Ltd, is a modern, 300 ton capacity vessel, 128ft long. It is seen here passing over the Stanley Ferry Aqueduct on the Wakefield section of the Aire & Calder Navigation. The skipper takes his car along so that he can go home in the evenings instead of cooking his own meals on board.

that were more square in section than previously. In the postwar years, economies of operation demanded larger inland waterway vessels. This was good for waterways that could accommodate larger craft, but on others it caused increased difficulties for the operators. The result was that the building of formerly common, but now small 'Sheffield' size vessels, 61 feet 6 inches long, became a rare event. The last small vessels were built during the early 1960s, while many other short vessels were cut in half, and a middle section inserted, and so lengthened to increase their carrying capacity.

Because of improvements made to the main line and Leeds branch of the Aire & Calder Navigation in the 1960s, 500 ton capacity oil tankers were able to deliver direct to Leeds in the late 1960s. The British Waterways Board, intent on trying to improve certain of their commercial waterways, enlarged many of the locks on the same navigation, and this allowed a barge with a 600 ton payload to travel to Leeds for the first time in 1978. The vessel, an oil tanker, is the first one of this increased size, but with all modern refinements included it cost in excess of £300,000 to build.

Despite the high costs involved a second one has been built and more large capacity craft are planned, especially now that the South Yorkshire water highway, the Sheffield & South Yorkshire Navigation, is to be modernised in the near future. This will allow craft with a 700 ton capacity access to Mexborough and 400 ton to Rotherham, and will replace the outdated and now uneconomical 'Sheffield' size steel-constructed Yorkshire keels with a cargo capacity of 90 to 100 tons, which have continued working into the 1970s.

Vessels of the Leeds & Liverpool Canal

THE LEEDS & LIVERPOOL CANAL, now the only trans-Pennine waterway open, was originally the first of three such waterway routes to be commenced, but because it took 46 years to complete, it became the last to be opened throughout. It had the longest main line of any canal in Great Britain, being 127 miles long from the River Aire to the River Mersey. In addition to a through-route, it provided a localised transport service for the many towns and cities on its course. Construction of the canal began at both ends, with the first section open in 1773 from Bingley to Skipton. Seven years later the canal was open from Leeds to Gargrave in the east and from Liverpool to Wigan in the west. These two sections enjoyed satisfactory independent trading for a further 36 years before they were joined together to become a through-route connecting the east with the west.

The canal had an average width of 45 feet and two tunnels, with, originally, nearly a 5 foot deep channel. It had two sizes of locks, 72 feet by 14 feet 4 inches from Wigan to Liverpool and 10 feet shorter from Wigan to Leeds. The Lancashire section incorporated the Douglas Navigation which had been completed in 1742, on which had operated Douglas flats, vessels 72 feet by 14 feet. The size of the locks resulted in the evolution of the Leeds & Liverpool 'short boat', 60 to 62 feet long with a 14 foot 3 inch beam. This type of craft, peculiar to the canal, was capable of carrying a little over 40 tons, at the rather shallow draught of 3 feet 9 inches. Of course in the Lancashire area, with the longer locks, vessels 10 feet longer worked locally, from Liverpool to Wigan and down to Manchester via the Leigh branch.

The building of these wooden vessels had obviously been influenced by the sailing vessels in operation at either end of the canal. The largest was capable of carrying up to 70 tons. The majority of the vessels had a square transom stern, with round bows, and were certainly not as bluff as the sailing vessels, and their stems had a little more shear. The hulls were of wood, mainly oak, with planking of carvel construction (edge to edge with caulking in between), fixed to a skeleton frame consisting of a backbone, the keel and keelson. For the ribs, the frames were usually fixed at right angles midships, while those located fore and aft, where the boat had pronounced shape, were made of carefully sawn and adzed pieces of oak. Those in the hold section were made of iron and wood.

Accommodation for the crew was provided in two cabins, positioned below deck level. Headroom being limited, it became the practice in later years to incorporate a low superstructure above part of the cabin. The aft cabin, which was by far the larger, was used to live in, while the fore cabin was used to store

29 *Mary*, an excellent example of a transom stern Leeds & Liverpool 'short' boat, was photographed at Hirst Wood, Shipley. Note the traditional decoration and fittings.

provender for the horse along with spare gear. Exception to this practice was when the family had many children, and then some of them would sleep there. Heating and cooking facilities in the cabins consisted of a coal-fired, black leaded stove. One type in common use was known as a 'pot bellied' stove because of its peculiar shape. The smoke from the stove above deck level went up through a tapered square-section chimney of wood.

The aft cabin was satisfactorily furnished with folding table, bench seats with lockers underneath, and at the sides, bed spaces. To complete the fittings, there were more cupboards and drawers for food and clothing, with a polished copper paraffin lamp for lighting at night. The drinking water was stored in a decorated barrel on the aft deck.

The boats had stern rails, but no hawse timbers at the bow, and were decorated in a unique manner, incorporating, especially on the transom stern, ornate scrolls and designs including bunches of fruit. Occasionally the gunnel, both fore and aft, was painted with a run of small connecting triangles of red, white and blue, commonly referred to as dragons' teeth. The hold, which was usually covered with a tarpaulin sheet, had a false bottom over the bilges. It was not very deep and had two cross members at deck level; these transverse deck beams were provided to reinforce the strength of the vessel. The base of the towing mast was positioned in the hold, approximately one-third of the way from the bow to the stern.

Leeds & Liverpool broad-beamed vessels also worked on the Lancaster Canal, mostly built by yards on the cross-country waterway. They were up to 72 feet long and carried in the region of 50 tons on a 3 foot 4 inch draught.

In later years similar shaped craft of iron and steel, still with a transom stern, were built by a yard on the River Ribble. These craft gained access to the Lancaster canal via Glasson Dock, as did the former wooden vessels. The final commercial traffic on the canal terminated in 1947, when a mill that had until then received supplies of coal for its boilers changed over to oil.

Horses, hauling from the towpath, were the accepted mode of propulsion until the 1880s. Then, because of the satisfactory operation of steam propulsion on certain neighbouring waterways, the directors of the Leeds & Liverpool's own carrying concern decided to try this new method. They commissioned G Wilkinson & Co Ltd of Wigan, who for a number of years had been building steam engines, to construct some engines for them, and thus a number of new vessels were built with steam engines installed. They were supplied with Edward Field vertical water tube boilers, which had been used most satisfactorily in fire engines, and were capable of raising steam quickly. The whole units, neat in appearance, were not as large as others that were being manufactured at the time, and took up little space in what, on other craft, was the aft living cabin.

30 These wooden vessels on the Lancaster Canal were very similar in construction to the Leeds & Liverpool vessels. The smartly dressed people on board must have just joined for a ride.

Decoration on the bow section of a Leeds & Liverpool Canal vessel

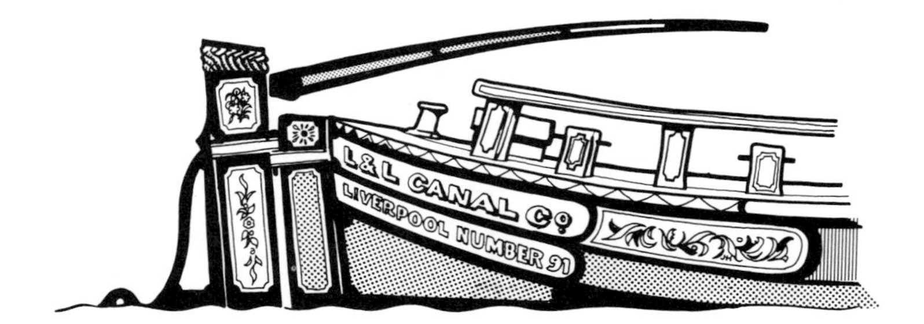

Decoration on the stern section of a round-sterned Leeds & Liverpool Canal vessel

Decoration on the stern section of a transom-sterned Leeds & Liverpool Canal vessel

The specially built steam boats were designed to haul a dumb boat, formerly a horse-drawn vessel. They were different from normal, in that they had a round stern to ease the flow of water to the 3 ft diameter propeller. From the very beginning the engines were economical in fuel consumption, burning either coal or coke, but over the years they were improved even further, until they burned only about 33 cwts of coke per 100 miles. A fireman was employed to attend to the boiler and engine, so the helmsman from his station at the tiller could easily operate the controls to stop, start or reverse the engine, blow the whistle or lift and lower the hinged funnel when required for low bridges.

The steamers were used initially on a regular 24-hour fly-boat service that proved most satisfactory, so the company continued to expand its fleet of steam-powered boats, commonly referred to as 'steamers'. The Wilkinson

31 A posed photograph, showing a Leeds & Liverpool round-sterned wooden boat; the transom stern was far more popular.

engine became standard, being fitted into numerous craft. Technically, it was a diagonal compound, non-condensing, double tandem engine. The engine had four cylinders, two being of low pressure and two of high pressure, and inclined in pairs to form a vee layout, with high pressure cylinders above low cylinders.

In the early 1920s the Leeds & Liverpool's own carrying concern was experiencing many problems, like many others at that time, so they terminated activities, selling the majority of the vessels, including the steamers, to bye-traders. In 1930, Canal Transport Ltd was formed, consisting of the Leeds & Liverpool Canal Company and four independent carrying concerns, namely B C Wells Ltd, Liverpool Wharfage Ltd, Lancashire Canal

32 A Leeds & Liverpool steamer is in the centre of the picture, while all around are wooden Leeds & Liverpool boats, both round and transom stern types.

Transport Ltd, and John Hunt (Leeds) Ltd. This became the largest operator on the canal. The new company, having money to spend on new vessels, commissioned a number to be built. The majority were steel boats with a new hull design, with a stern section that was almost vertical above the waterline and with a nearly semicircular aft deck section, like the bows.

The new vessels, both motorised and dumb, were mostly named after rivers. Because steam engines were labour intensive, the motorised vessels were fitted with 24 hp Widdop diesel engines, made in Keighley. Many of the new vessels, built by Isaac Pimblott & Son Ltd of Northwich, were 61 feet in length, with a beam of 14 feet and with a loaded draught of 4 feet 8 inches.

Meanwhile, other operators continued in the old manner with horse-drawn wooden vessels, although some bought new motorised vessels of both wood and steel. The steamers owned by the bye-traders continued working satisfactorily. Some of the older vessels were wearing out, but because the machinery was in such good condition it was installed in new hulls. Only in the mid 1940s did the steamers start to lose favour, and only then because of an accident, when a crewman was killed by a flying rivet from a boiler. The insurance companies insisted on expensive modifications which included the replacement of all the boiler rivets with specially-made bolt-type fittings. This was considered by many as too expensive, as the majority of the vessels and the machinery were old. Yet over a dozen boilers were modified to comply with the new specifications, and these went on operating for another decade or so.

Unfortunately from the mid 1950s, traffic continued to decline alarmingly, and vessels continued to be withdrawn until the last regular working in 1973, when the coal supplies to the Wigan Power Station in steel motor boats ended. A few Leeds & Liverpool boats still remain, both wooden and steel. The steel ones are owned either by enthusiasts, or keen supporters of canal transport like Northern Counties Carrying Ltd and Apollo Canal Carriers Ltd, who have spent a great deal of effort to secure a little work, or by the British Waterways Board which uses them for maintenance work on the canal.

33 A former steel motor 'short boat' from the fleet of Canal Transport Ltd, of the Leeds & Liverpool Canal, receives a transhipped cargo from the 200 ton capacity, steel dumb barge moored alongside the Aire & Calder Navigation wharf at Leeds in 1949.

34 Bomb damage on the Leeds & Liverpool Canal at Liverpool.

Inland Waterway Tugs

When the navigations and canals were constructed in the first instance, there was no mechanical power available, or even contemplated, that would propel boats along the waterways. Therefore practically every mile of navigable waterway had a hauling way, or towpath, provided. For many years, when the wind would not fill the sails, man hauled vessels on certain waterways before animals such as horses, mules, donkeys – and even bullocks on occasions – became the accepted method of propulsion. Men and animals need rest after strenuous effort, and even nature, harnessed by the use of sails, was not always obliging. So one would think that attempts to overcome the problems by the use of mechanical power would have been eagerly encouraged when it first became available. Incredible as it may seem now, when a steam engine was fitted and used to propel a vessel along an inland waterway, it was practically condemned, and considered far from suitable.

The first recorded attempt at using mechanical power on an inland waterway was as early as 1797 on the Bridgewater Canal. The most renowned test, however, was in 1802, when the *Charlotte Dundas*, a centre stern paddle steamer, was used on the Forth & Clyde Canal to haul two vessels loaded with about 70 tons of cargo, at an average speed of 2½ miles an hour. The trials were successful, as far as the vessel's operation was concerned, but the critics were not impressed, saying that the vessel's wash would damage the canal banks, while silting would occur to impede the progress of other vessels. So convincing were these and other criticisms against further attempts to provide a satisfactory steam propelled vessel, that many years were to elapse before serious consideration was to be given to the potential of this revolutionary method by the canal companies.

Despite these initial criticisms, the steam tug did have a few supporters, who saw a great future for a small, powerful, self-propelled vessel, which could haul a number of vessels efficiently along a waterway completely independent of horses and wind. The only problem was to convince the actual users, but eventually they succeeded. One of the first inland waterways to introduce a steam tug onto their network was the Aire & Calder Navigation, which took delivery of the first tug in 1831. The vessel was put to work on a section of waterway where there were few locks, hauling fly-boats between Castleford and Goole. It was a great success, and a second vessel was ordered almost immediately. Two years later, in 1833, a director of the company reported: 'Steam towing is attended with advantage in speed and economy.' These two tugs were the first of many that the navigation company and its bye-traders

35 For many years there were thousands of horses treading the towpaths of Britain – even as late as the early 50s there were still several hundred at daily work. The harness includes painted wooden bobbins to prevent chafing of the animal's flanks.

were to operate for over a century. They were used to haul many millions of tons of cargo, the navigation company even operating a towing service for bye-traders to a timetable, in addition to operating a fleet of tugs especially for hauling the Tom Puddings, the company's compartment boats.

The success of the steam tug *Birmingham*, which in 1855 hauled 20 barges with a gross weight of 1,231 tons on the Regent's Canal, convinced many that steam-powered tugs were here to stay. It was not long afterwards that their use became widespread. Tugs were especially useful on river navigations and canals with long lock-free pounds. On the Regent's Canal they were used extensively, working to and from the Limehouse docks, while in the same area they were to be found on the River Lee, on the lower Grand Junction Canal and, of course, on the River Thames.

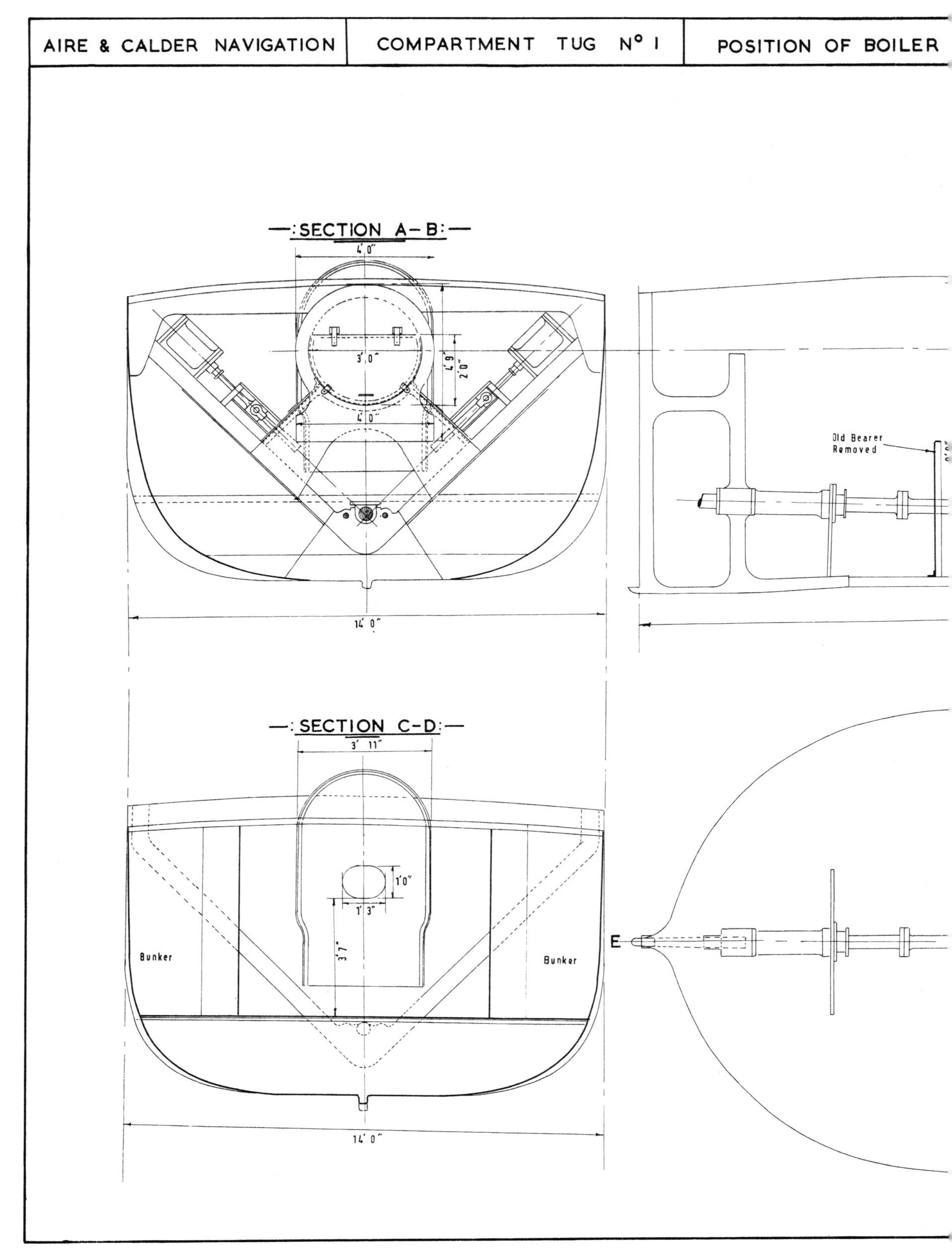
AIRE & CALDER NAVIGATION
COMPARTMENT TUG N° 1
POSITION OF BOILER
—:SECTION A–B:—
4' 0"
3' 0"
4' 9"
2' 0"
4' 0"
14' 0"
Old Bearer Removed
—:SECTION C–D:—
3' 11"
1' 0"
1' 3"
3' 7"
Bunker
Bunker
14' 0"
E

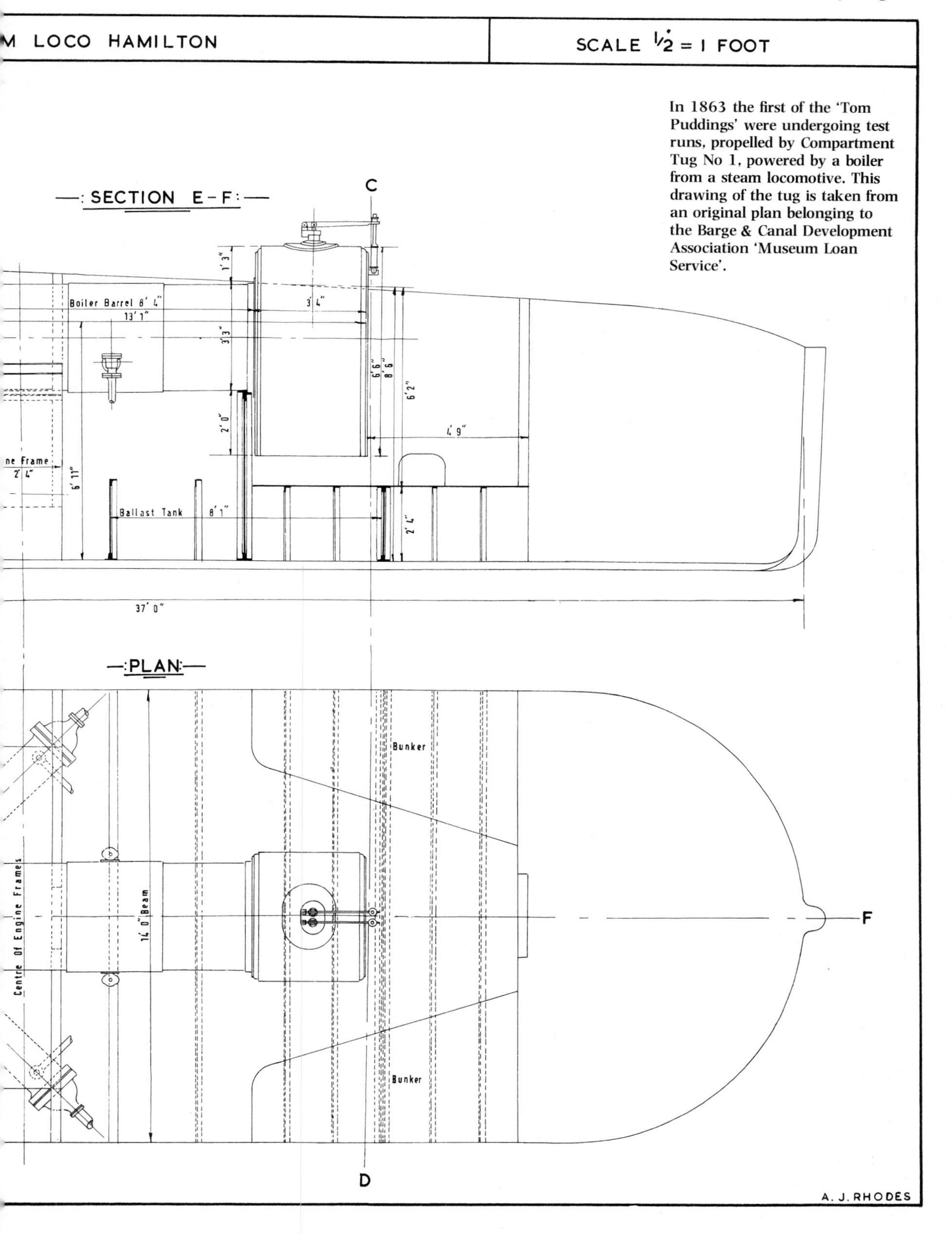

In 1863 the first of the 'Tom Puddings' were undergoing test runs, propelled by Compartment Tug No 1, powered by a boiler from a steam locomotive. This drawing of the tug is taken from an original plan belonging to the Barge & Canal Development Association 'Museum Loan Service'.

36 Narrowboat steam tug *Antelope* hauls numerous vessels both loaded and empty in 1911 at Alperton.

They were widely used on the Gloucester and Sharpness Canal, and on the local River Severn, where they were invaluable, regularly hauling many and varied craft. Further north, despite having a heavy locked waterway on which they operated steam packets, the Rochdale Canal owned a number of tugs that operated on certain local waterways, like the River Weaver and the Bridgewater Canal. Those on the Bridgewater were used for hauling dumb vessels between Runcorn and Rochdale, taking 21 hours for the 43 mile journey. The tugs never went beyond Rochdale as the canal was too shallow to take the 4 foot 2 inch diameter propellers that turned at an average speed of 180 revolutions per minute.

In the 1870s the Bridgewater Navigation Company ordered 26 tugs, after using one on test for a while. They were small and elegant, with tall funnels and long, low superstructures. Based at Runcorn, they were used extensively for over 70 years, being known locally as 'little packets'; most of them were eventually dieselised. They hauled up to three loaded flats and other unpowered craft belonging to the company and bye-traders. The Leeds & Liverpool Canal Company owned a few tugs, which were used in conjunc-

37 On the Bridgewater Canal a steam tug hauls two empty general cargo 'Dukes' to Runcorn.

tion with horses. The horses hauled the fly-boats along short pounds and through the locks to the waiting tugs which then hauled along the lengthy pounds. The tugs usually remained on their allocated section of canal.

Over the years the tugs were used increasingly on the rivers with inland waterway connections, like the Mersey, Humber, Trent and Severn. Where the rivers were wide enough, the towed vessels would be fanned out behind the tug, and individual vessels which were not travelling the whole distance were located at the back, so that they could slip their towing lines at their required destination, whilst the tug and the rest of the flotilla continued on their way without stopping.

A number of canal tunnels had been built without towpaths, and this resulted in boats being propelled through by the back-breaking method of 'legging', a slow and lengthy process which caused many delays and hold ups. Narrow planks of wood, known as 'wings', were laid across the boat, and the men would lie on their backs along the planks and 'walk' along the tunnel walls, pushing the boat through. Meanwhile the women and children would lead the horse over the tunnel to join the towpath on the other side.

38 Again on the Bridgewater Canal a 'Duke', 71ft long 'flat', is discharged, alongside *Stretford*, one of the company's steam packets.

In 1871 the Grand Junction Canal Company introduced steam tugs for hauling vessels through Blisworth and Braunston tunnels, which considerably speeded up the movement of boats. The idea caught on, and later steam tugs were used on the Trent & Mersey canal at the tunnels at Preston Brook, Barnton and Saltersford; on the Leeds & Liverpool Canal at Foulridge and Gannow; at Islington on the Regent's Canal; and on the Worcester & Birmingham Canal at Tardebigge, Shortwood and Westwell. Some of the tunnel tugs were not orthodox in appearance. Some were double ended, and others had wheel fenders, provided to run along the tunnel walls. The tugs continued in use until the majority of the vessels passing through the tunnels were self-propelled. Then they were withdrawn, in most cases during the 1930s.

The last tunnel tug in use was at Harecastle on the Trent & Mersey. Here an electric tug, originally powered by batteries and later by mains electricity, had been provided in 1914, to be used instead of steam powered tugs because of the very restricting headroom of the tunnel as well as the poor ventilation. It was withdrawn from service in 1954.

39 A former Leeds & Liverpool Canal Company wooden constructed steam tug, in the early days of nationalisation, is being used on the canal to break ice. Note at the bows an attachment used to protect the vessel.

57

40 Steam tug *Olga*, owned by the West Norfolk Farmers Manure Company of Kings Lynn, is seen here waiting at Bottisham Lock on the River Cam, while her tank barges are locked through.

Two diesel powered tugs that had been employed as tunnel tugs still exist, both preserved; they are the *Worcester* and *Sharpness*. Constructed by a shipbuilder, they both have very pronounced counter stern sections, along with wheel steering. Built between 1908 and 1912, these iron-hulled craft, 45 feet long, look very much like small river tugs.

After the turn of the century, the internal combustion engine was increasingly fitted into tugs, as well as boats and barges. The petrol and paraffin, and later diesel engines, were not as labour intensive and came into common use, although many operators remained loyal to steam power. Some steam tugs were built as late as the 1940s and continued in use until the 1960s. Indeed a few of these, converted from steam to diesel, are still in use today.

Narrow-beamed diesel tugs, some full 70 feet long and others about 40 feet overall, were certainly very popular in the Birmingham area. There several dozen of the powerful, yet handy, tugs worked for many years hauling strings of narrowboats between the collieries and the wharves of gas works, electricity undertakings and industrial premises; many of them remained at work until the late 1960s. Of all the tugs built over the years, there are more ex-Birmingham-area ones in existence today than any other type, although these are now owned by private enthusiasts rather than commercial companies.

Tunnel tug

41 *Else Margareta*, a diesel powered tug owned by Hargreaves (West Riding) Ltd, hauls large steel dumb barges on the lower section of the Calder & Hebble Navigation at Wakefield in the early 1950s.

42 *Audrey*, a powerful steam tug owned by J Hargreaves (Leeds) Ltd, was for many years employed hauling dumb barges loaded with coal to works and power stations. The introduction of motor barges in the early 60s made the vessel obsolete and it was sold for scrap.

43 This diesel-powered narrowboat tug, still owned and operated by a Birmingham Company, is now used mostly for moving day boats on salvage and rubbish collection.

44 Very few tugs are now owned by private operators in the north east; one of them is *Cawood*, owned by John Whitaker (Holdings) Ltd. It was built in 1912 as a steam powered vessel; later converted to diesel, it is now used only occasionally.

45 These are two of the British Waterways 'Tom Pudding' tugs, part of a new fleet built in the late 50s and early 60s to replace the former Aire & Calder Navigation steam tugs. They were named after West Yorkshire collieries; the vessel near to the wharf has a Jebus fastened to its bows.

On the waterways where commercial carrying still exists, very few tugs are now employed towing conventional dumb barges, because very few dumb craft remain. The majority of the working tugs are employed on maintenance work, or in the North East towing trains of Tom Puddings. In the 1960s large capacity compartment boats were introduced, providing a modern water transport system. To move these compartment boats, a modern tug that was exceptionally easy to manoeuvre was introduced. These 'push tugs', being rectangular shaped craft fitted with powerful engines, have directional drive units that enable them to turn in their own length. They usually work with three compartment boats that have a cargo capacity in the region of 500 tons.

Containers and Compartment Boats

CONTAINERISATION was in use on the inland waterways two hundred years ago. The renowned James Brindley designed coal containers to transport coal on the Bridgewater Canal, and similar ones were used later on the local waterways of the Leigh branch of the Leeds & Liverpool Canal and the Manchester, Bolton & Bury Canal. They were small wooden containers 6 feet by 4 feet 6 inches by 4 feet deep, with a capacity of 35 cwt. They were carried ten at a time, from the colliery wharves to their destinations in special boats 68 feet long with a 7 foot beam. At the consignee's wharf they were craned out of the boats and the contents discharged through doors in the bottom. They continued in use until just a little over a decade ago.

A canal from Framilode on the River Severn to near Stonehouse used containers that carried about 1 ton each. They were transported in specially designed craft that carried about 8 of these wooden box containers. Instead of using locks on the canal, cranes were located at every change in level to transfer the boxes to another boat in the next pound. These boats operated on each independent canal pound only.

From the mid 1790's at Denby colliery near Derby, coal was loaded into wooden containers. They were carried on a waggonway in trucks, hauled by teams of horses, to Little Heaton at the head of the Derby Canal. There, by means of a crane, they were transferred to canal boats, which usually carried five of the containers at a time. The practice ended in 1908.

British Transport Waterways, the forerunner of the British Waterways Board, introduced containers in the 1950s in an endeavour to obtain additional traffic. One type, constructed of reinforced glass fibre, had metal bands around the exterior to assist lifting. Two sizes were introduced, one 7 feet high and 6 feet square, capable of carrying 4 tons, and a smaller one 5 feet high and 5 feet square. The large ones were used on the North East waterways for exporting goods from inland terminals at Leeds, Sheffield and Nottingham, and were transported in the holds of barges to the Humber ports for shipment overseas. One regular service carried washing machines for export and returned with imported Dutch bulbs. The smaller containers, designed for movement by narrowboats, were used for some years on the Grand Union Canal, being transhipped at Hayes Wharf. In addition to the fibreglass containers, they also introduced aluminium open-topped containers, designed to fit on the back of a lorry and to stack in the holds of barges. Unfortunately because of industrial troubles, all the containers, which had proved to be highly satisfactory in use, were withdrawn after only a short period.

46 On the Bridgewater Canal, the tug *Phyllis* with the loaded wooden barges astern, is passed by a considerable number of the container-type narrowboats loaded with the coal containers from the Leigh branch of the Leeds & Liverpool Canal.

Compartment boats, rectangular floating boxes, are practically as old as the canal system itself. They were used on numerous canals from Devon and the South West to Wales and Shropshire. Usually small in size, and fastened together in a train, they were hauled by men and horses. On the Marquess and Stafford Canal of the Shropshire Union Canal Company's connected canals, they were generally of one basic specification, 20 feet long by 6 feet 4 inches broad, and square ended. They were in use at Donnington Wood from about 1786. Their basic design set the pattern which later canal companies in the district were to follow.

Tub boats, as the early compartment boats were known, were a cheap and convenient method of transporting materials of all kinds. On occasions over remarkably short distances of a mile or less, they were a far more economical form of transport than roads or tramways and in the early days of iron making many works owned and operated boats. For example, in 1800 the Ketley Iron Works owned over 200, while the famous Coalbrookdale Company bought 29 in January 1805, costing less than £20 each.

One horse, hauling any number between 12 and 20 boats, was controlled by a steersman walking on the bank. At many places in the Shropshire area, tub boats worked to transhipment points like Trench, Wappenshall and Lubstree

47 One of the British Waterways' fibreglass containers, 5ft square and 6ft high, introduced in the 1950s for movement by narrowboat. This one is owned by the Barge & Canal Development Association, Museum Loan Service.

for national distribution. The majority of the tub boats continued in use until the 1840s when the use of the narrowboat led to their decline. On different tub boat canals they had different carrying capacities, ranging in size from the smallest on the Donnington Wood Canal with 3 tons, to the Shropshire Canal with 5 to 6 tons and on to the Ketley and Shrewsbury Canals with 8 tons. Tub boats carrying 5 tons were still in use during the First World War on the Coalport Canal in Shropshire.

In Devon on the Torrington Canal the boats worked in pairs, the first with pointed bows, close coupled to a second, square-ended boat. One of the most successful tub boat canals was the Bude Canal, opened in 1819. It was used mainly for conveying sea-sand, which was used as a fertiliser, to the farmsteads above Bude. In its heyday the canal stretched 46 miles and had six inclined planes. These were used instead of locks for lifting and lowering the boats between pounds. Some of the inclined planes were real feats of ingenuity, as was the Habbacott Incline, 900 feet long and with a vertical height of 250 feet. To lift the boats containing about 4 tons of sand, an enormous bucket was

48 This 20ft by 6ft by 4ft wrought-iron tub boat from Lilleshall is of early nineteenth century construction, and is typical of the tub boats used on the Shropshire Tub Boat Canal.

49 The simply constructed Bude Canal compartment boat.

filled with about 15 tons of water, which was duly lowered down a 250 foot shaft. This pulled the boat upwards, but only part way. Then a second bucket of water, lowered down a second shaft, completed the lift. The tub boats were of two different shapes and sizes; the majority had square ends, were 20 feet in length and 5 feet 3 inches wide, and worked to a loaded draught of 1 foot 8 inches. A number, used as the leading boats, were provided with pointed bows, and were 3 feet longer. All of the boats were fitted with wheels, which were used for transporting the boats up and down the inclined planes.

50 At Castleford, on the Aire & Calder Navigation before nationalisation, a steam tug pushes the Jebus in front while towing empty Tom Puddings; on the loaded train the Jebus can be seen to really do its job of splitting the back stream from the tug's propeller.

In 1860 the Aire & Calder Navigation hoped to transport vast amounts of coal from collieries in the western region of their network to Goole. They were looking for an efficient system to compete with the railways, and W H Bartholomew, their engineer, tried to solve the problem by considering tub boats. He proposed a system with boats larger than had been used before, capable of transporting about 25 tons and hauled by tugs, instead of horses. To speed up the turn round at Goole he designed a hydraulic hoist that would lift the loaded compartment boats out of the water and tip the contents directly into the holds of sea-going vessels. Initially it was planned to operate the compartment boats in units of six with false bow and stern units from where the steering would be done.

The Navigation directors, assured of Bartholomew's expertise, gave the go ahead for the system, which, by late 1864, proved to be an efficient method of transporting coal. It was found that when the compartment boats were secured together with chains as a rigid unit, there was no need for a stern unit. To the directors' satisfaction it was also found that it was possible to haul far more than six units at a time. Small modifications in the system continued to be made and it was duly patented, with Bartholomew receiving a satisfactory royalty for every ton transported. With continual improvements, they were eventually made of iron, 20 feet long by 15 feet wide, capable of transporting 35 to 40 tons each, and towed by steam tugs in trains of up to 20 compartments at a time.

In front of the leading loaded 'Tom Pudding', it became standard practice to have a false bow unit, which was also used as a towing piece, known as a 'Jebus'. The Jebus's main function was and still is to separate the water pushed back from the tug's propeller, to prevent the water building up against the first loaded Tom Pudding. As this has a 6 foot draught it causes adverse effects on the handling and propulsion of the train. When the tug is hauling empty compartment boats, the Jebus is not required, for the backwash from the tug mostly flows underneath the train, and so it is usually positioned in front of the tug's bows.

Coal was normally loaded by gravity into the compartment boats, or pans, as they are sometimes referred to, at waterside colliery staithes in the Wakefield and Castleford districts, and after 1930 also from Hatfield colliery on the Stainforth & Keadby Canal, via the New Junction Canal. One exception to the normal practice of loading was at Newlands colliery, near Wakefield. There the boats were loaded directly under the colliery screens, and travelled to and from the navigation on specially built low slung bogies, hauled by a steam locomotive. This unusual practice continued until 1942. At Goole five hoists, capable of lifting 52 tons (i.e. more than the combined loaded weight of boat and coal), were constructed, one of them mobile. Over the years three of the hoists have been dismantled because of old age, leaving two still in daily use.

In 1914 there were 1000 compartment boats in use and many millions of tons of coal were transported by this efficient and economical method. But since then there has been a slow but continual reduction in their numbers, due in part to a decline in coal exports. In fact since the mid 1970s no coal has been transported by the system, the tonnage being replaced in part by smokeless fuel. The tonnage shipped now totals in the region of 200,000 tons annually, transported in the remaining 299 compartment boats. They are now hauled by four diesel-powered tugs that replaced the former steam tugs in the late 1950s.

In 1967 a new revolutionary system of inland waterway transport was inaugurated: the use of new compartment boats 56 feet long by 17 feet 3

51 *C.H.103*, one of the Cawood Hargreaves Ltd 'push' tugs, hauls three of their 165 ton capacity containers to the Savile colliery for loading, seen from the cab of a sister tug pushing three loaded compartment boats.

inches wide, with a carrying capacity of 165 tons. Usually three at a time are pushed along by the new 'Push Tugs'. The system incorporates a hoist that takes 9 minutes to unload a compartment boat. Since its inauguration, the new push tugs and compartment boats system has carried in excess of one million tons of coal every year from West Yorkshire collieries to Ferrybridge 'C' power station on the Aire & Calder Navigation. The 9 push tugs and 35 compartment boats in use are owned and operated by Cawood Hargreaves Ltd.

British Waterways Board, aware of the tremendous possibilities of push tugs and compartment boats, built two tugs at their Goole yard. They are *Freight Pioneer* launched in 1970 and *Freight Trader* launched in 1971, and they work with 9 compartment boats 55 feet long by 15 feet wide which the Board had built. Each of these compartment boats, with hold covers, is capable of transporting 140 tons of general cargo. They were used initially on the Sheffield and South Yorkshire Navigation, as it was necessary to replace their aged steel motorised keels, which were nearly 50 years old.

A private company, Bacat Ltd, saw the tremendous potential in a

52 *Freight Trader* one of the British Waterways Board 'push/pull' tugs with three empty compartment boats on the Sheffield & South Yorkshire Navigation.

compartment boat system – especially if the containers could be loaded inland, moved on the inland waterways to the Humber estuary, and there be taken aboard a 'mother ship' to be taken abroad without any need for transhipment. As a result the company built 63 compartment boats, of the same size as those of the Board, along with a 'mother ship'. The ship was smaller but similar to others in use throughout the world that transport LASH and SEEBEE containers, which are too large for the inland waterways of Britain. Yet the operations of Bacat Ltd were withdrawn after only a short time. Although it proved to be a satisfactory method, the service was terminated because of industrial troubles. The British Waterways Board, having purchased additional tugs to propel the Bacat barges on the inland waterways of the North East, have recently purchased 13 of the former Bacat barges to add to their own fleet. These compartment boats are now commonly referred to as 'Rotherham' barges, and are now used for transporting general cargo to and from the Humber ports to inland terminals, both private and those operated by the Board.

Vessels of the North West

vicinity of the Mersey estuary, the vessel that had evolved from many centuries of local trading was the Mersey flat. It was not of a standard size, but around 65 feet long by 16 feet beam, with a distinctive pointed stern. Its trading area was around the local coast, up the River Mersey to Runcorn and Warrington, as well as on the local River Weaver, to the tidal limits at Pickering, a little above Frodsham.

Owing to an Act of Parliament in 1721 that encouraged improvements to local rivers, in 1732, for the first time, sailing vessels were able to proceed upstream on the River Weaver to Northwich and Winsford. The operating area of the flats was further extended in 1736 when the Mersey & Irwell Navigation was opened to Manchester, and it was extended even further in 1757 with the opening of the Sankey Brook Navigation to St Helens. This waterway had locks 68 feet long by 16 feet 9 inches wide, allowing access to vessels loaded to a draught of 4 feet 6 inches.

The size of lock on the Sankey Brook Navigation gives some indication of some of the size limits of the Mersey flats at that time. Due to the fact that the flats were increasingly working through locks, the traditional pointed stern was gradually replaced on new vessels by a nearly square stern, which eventually became standard. Also, from about that period, the rigging on new vessels became a standard gaff, fore and aft rig, usually a peaked version, on a single mast. This was a convenient rig for a crew of two to operate.

A further expansion of the operating area came when the Bridgewater Canal was extended west to Runcorn in 1776, the locks there allowing vessels 72 feet long by 14 feet 2 inches wide to pass. The locks only allowed access to the canal to the narrowest of the flats then in common use. Additional waterways in the area were opened, increasing yet again the navigable area. Such a large area of operation for wide-beamed vessels resulted in many variations in vessel size and shape, for many modifications were incorporated in new vessels which were built for internal canal use only. Some of the craft had both full round bows and stern, and were of a much shallower draught than formerly, but all were based on the accepted traditional building practices of the region.

In 1804 the Rochdale Canal was completed, many miles inland from the Mersey. It had locks 74 feet long by 14 feet 2 inches wide to accept the standard, small Mersey and Weaver flats then being constructed. These vessels then, for the first time, were able to travel over the Pennines into Yorkshire, but were unable to go any further east than Sowerby Bridge on the next section of the first cross-Pennine waterway link, because of the restricting Calder &

53 A 'flat' in full sail on the River Mersey in 1892.

Hebble Navigation locks, which were only 57 feet 6 inches in length.

On the canals the flats relied upon horses hauling from the towpath, while on the tidal rivers they used sails. In the 1830s the Mersey & Irwell Navigation's engineer, Samuel Wylde, designed a flat to be constructed of iron. A few were built and found to be most satisfactory. They were the exception, for timber continued to remain the accepted building material. From the mid 1800s tugs became popular to haul vessels on the rivers. The proprietors of the Mersey and Irwell Navigation were the first to unrig their sailing flats in favour of propulsion by steam haulage on the River Mersey.

By the mid-nineteenth century, two standard types of flats had evolved: the dumb vessel, hauled along the inland waterways, and a sailing vessel of a much larger size, acting as a short distance coaster. Both types of vessel had a hull with a flat bottom and a rounded bilge, as well as bluff looking bows. As the sailing vessels did not have to negotiate locks, once more they appeared with a pointed stern, the fine run aft, starting from the region of the hold aft bulkhead. With two fixed masts, they were jigger rigged, the former square sails being replaced by rounded ones. The running rigging was of wire, while the blocks and large free running sheaves were of wood. The other ropes and shrouds were of hemp. For practical reasons the sails were usually dyed a brown colour.

The large sailing flats usually had a crew of three who lived in a cabin at the stern, below decks. Access to this was via a sliding hatch, the berths were of the recessed type, fitting along the sides of the cabin, and at the side of the bulkhead, near the cargo hold, was the coal stove. Lockers and cupboards for clothes and

54 The Rochdale Canal Company owned a large fleet of 'flats' mostly named after flowers and trees. Here *Rose* is being unloaded at Sowerby Bridge early this century.

food were around the stern section of the cabin. The vessels, of carvel construction, were built of oak, with large sectional frames and thick planking. Unlike many sailing barges, leeboards were not used. Instead they relied upon the vessel's external keel and the well shaped bow and stern to avoid side slip. The steering, in common with most other vessels at that time, was by means of a long, heavy sectioned tiller, giving the helmsman plenty of leverage to move the rudder.

The maximum tonnage carried by the vessels on the Mersey estuary was usually limited to about 170 tons, and this ensured that the draught did not exceed 9 feet. The cargo was stored in two holds, one fore and one aft of the main mast. On the fore deck were two large winches, the larger one used for lifting the anchor and the smaller one for handling the barrel line. This rope measured about 120 fathoms and was used for warping about the docks, and other hauling purposes. At the bows above deck level was the small but elegant hawse timber, with ample timber heads on either side for securing mooring lines and tow ropes. A store was provided forward in the bow section for spare ropes, tar and paint etc. A low rail ran on either side of the hull alongside the sidedecks leading to the after deck, where the drinking water storage barrel was to be found.

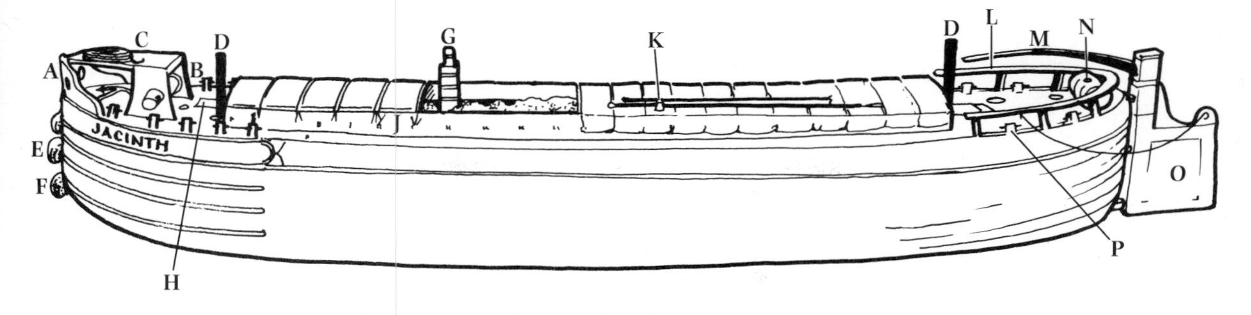

Rochdale Canal Company flat

Key:
A Hawse timber
B Winch
C Rope board
D Stove chimney
E Stem fender
F Cill fender
G Towing mast (lutchet)
H Hatch to cabin
J Coamings
K Boathook and stower
L Stern Rail
M Tiller
N Water barrel
O Rudder
P Timber heads

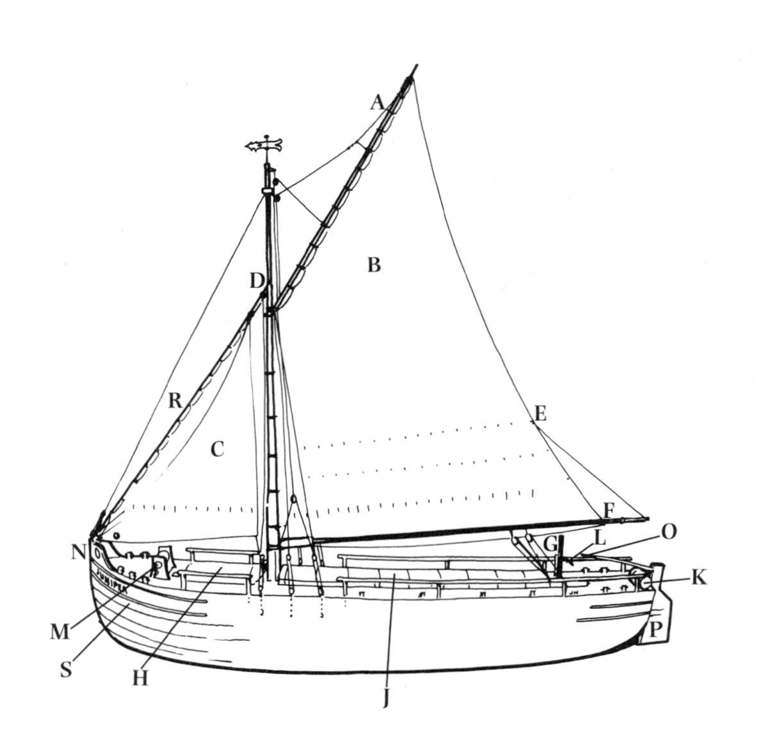

Mersey (single mast) sailing flat, 50 to 75ft long

Key:
A Gaff
B Mainsail
C Foresail
D Throat halyards
E Reef point
F Boom
G Stove chimney
H Forehatch
J Mainhatch
K Water barrel
L Pump
M Windlass
N Hawse timber
O Tiller
P Rudder
R Forestay
S Top strake

Decoration on these vessels was very simple. The hull was tarred up to its top plank, while the top plank and the hawse timber, wooden-framed cabin skylight and other deck fittings were painted in bright colours, and the mast heads on occasions were picked out in contrasting colours. The barge's name was usually carried on the top strake of the hull at both the fore and aft ends.

The Duke of Bridgewater's original waterway undertakings were the underground canals at the Worsley collieries, followed by the Bridgewater Canal. Narrowboats were always owned by the company and, from the earliest days, so were a considerable number of flats. In the 1890s the number increased, until they had a maximum of about 1000 flats. These were of three

55 Mersey 'flats' in 1892 at Princess Dock, Liverpool. The one nearest the camera, formerly a sailing vessel, has been de-rigged, but the one partly obscured shows the rigging clearly.

basic sizes. The largest, fitted with sails for river work, was unable to proceed through the Runcorn locks. The other two sizes were the same in overall dimensions, of 71 feet long and 14 feet 3 inches beam, the difference being in the draught. The 71 feet long vessels were known as Dukes, and while the deep draughted vessels could work through the Runcorn locks, they were unable to travel any further along the canal than Preston Brook Junction. The shallower draughted vessels worked throughout the system, carrying about 50 tons. After the 1870s the company's vessels on the canal were always hauled, usually three at a time, by small powerful steam tugs.

The traffic on the River Weaver continued to increase over the years, the biggest percentage of goods carried being coal upstream and salt down. Under good management the waterway continued to be improved, and in the 1860s they introduced steam tugs to haul dumb flats, which, over the years, continued to be built larger and larger. During the 1870s and 90s they modernised the system, building locks 229 feet long by 42 feet 6 inches wide. Then steam cargo vessels, always referred to on the River Weaver as 'packets', were introduced. These packets were employed working from Winsford and Northwich to the Mersey ports. The average packet was some 90 feet long and 21 feet wide, and it was not uncommon for them to tow three dumb vessels, which made a 1000 ton cargo movement in one unit. By the end of the nineteenth century, steam packets built of steel dominated the scene on the navigation. The last timber-constructed steam packet, which was built in 1897, was 121 feet in length and was named *Monarch*, while the first motor-propelled vessel to be introduced on the River Weaver was *Egbert* in 1911, capable of carrying 130 tons.

In 1935 only about 20 fully rigged Mersey flats remained, and these disappeared gradually until there was only one under sail on the River Mersey in the mid 1940s: *Keskadale*, owned by Richard Abel of Liverpool. Meanwhile, on the Bridgewater Canal, the depression of the 1930s had caused the number of company-owned vessels to be reduced to less than 80. In later years however there was an increase in the number of Leeds & Liverpool boats in operation.

The Bridgewater Company modernised their fleet in the late 1940s, and obtained a number of steel vessels, both motor and dumb. These vessels were built for them by yards on the River Weaver, and were of welded steel construction. The motor barges, with 68 bhp Gardner diesel engines installed, were named after Cheshire and Shropshire meres. These motor barges were able to carry 80 tons on a 5 foot draught, while the dumb barges on the same draught carried 114 tons. These vessels, along with a few owned by a couple of bye-traders, were really the only modern commercial vessels operating on the canal. They went on working for the next couple of decades, until the last commercial working on the Bridgewater came to an end in 1974, 200 years after it had all started.

On the River Weaver the last three steam packets were built in 1945/46. But, by the end of the 1960s, after 100 years of operation, all the steamers had gone from the navigation. During 1947/48, however, seven motor vessels were built for ICI Ltd for operation on the navigation. These motor barges were

56 *Syria*, seen here at Sutton Bridge on the River Weaver in August 1957, was then owned by ICI Ltd. Built in 1885 at Winsford, 88ft long, she was fitted with one pair of vertical engines manufactured by W E Bates of Northwich. In the background is *Osmium*.

57 The last mineral traffic on the Bridgewater Canal was coal to Trafford Park Power Station, which terminated in the mid 60s. The vessels that were employed on the job were Leeds & Liverpool Canal type motor 'short boats' and their Lancashire sisters which were about 10ft longer.

of similar design to the steamers, and were built locally, five from Yarwood and two from Pimblott of Northwich. In 1956, the British Transport Commission authorised a development scheme for the Weaver Navigation. The result was the elimination of Sutton lock and an improved channel, allowing 600 ton capacity coasters access for the first time. ICI Ltd now operate five 300-ton-capacity motor barges: *Comberbach, Cuddington* and *James Jackson Grundy* built in 1947, and *Marston* and *Marbury* built in 1948. In addition there are a number of bye-traders' vessels operating from ICI Ltd's wharves at Winnington and Wallerscote works and from the wharves of the British Waterways Board at Anderton. Coasters also work from the same wharves to Scandinavian, Scottish and Irish ports.

58 Seen here on the River Weaver is *Dovedale*, a former Bridgewater Company 'packet' obtained by Richard Abel in 1947, when it was renamed from *Lymm*. The vessel had been converted to diesel power before the sale; it continued in work until the early 60s when it was scrapped. The vessel in tow is a large capacity steel-constructed dumb vessel owned by the same operator.

59 The last regular traffic on the Bridgewater Canal was grain to this Kellogg's mill up until 1974. The vessels were owned by the Bridgewater Canal Company. In front is *Paradine*, a steel motor barge, powered by a 68bhp Gardner engine and capable of transporting 80 tons. To the rear are dumb vessels, hauled by the motor barges, capable of carrying 114 tons.

60 *Agnes*, a bye-traders motor vessel, is moored at the ICI wharves on the River Weaver. An old vessel, formerly steam powered, it is still in daily use.

Vessels of the South West

THE SOUTH WESTERN region of Britain's inland water transport is based on the River Severn and adjoining Bristol Channel. The vessel that dominated the scene for several centuries was the trow (pronounced to rhyme with 'throw'). The trow's area of operation initially extended upstream to the Severn's upper navigational limits, a little above Stourport, and down to the Bristol Channel ports.

In the mid-fifteenth century, when the river was becoming a main artery of early water transport, the common sailing vessel in use, referred to as an early trow, had developed into a double-ended vessel with square sails. Over the next couple of centuries, the sailing vessel gradually changed to the distinctive shape by which trows are generally known, with a flat bottom, pointed bows that had a slight shear, and a 'D'-shaped transom stern. The rigging then in common use was fore and aft ketch, cutter or sloop rig. No standard size of craft existed, so they varied considerably. The largest vessels were 70 feet long by 17 feet wide; working at a draught of 9 feet they were able to carry 120 tons.

During the late eighteenth and early nineteenth centuries improvements to local river navigations and the construction of canals in the area of the River Severn resulted in the trow's area of operation being extended considerably. The waterways that were used by the small trows and their variants included the River Wye, the Warwickshire Avon, the Kennet & Avon Canal and its river sections, the Thames & Severn Canal as far as Brimscombe, as well as the Droitwich and the Stroudswater Canals.

Trows had been employed from about 1580 to transport cargoes of grain, brought overland from the Midlands to the upper River Severn and then by water on to Gloucester. For many years agricultural products were the mainstay of traffic for the smaller vessels that operated on the rivers but did not make journeys on estuarial waters. After the opening of the Droitwich Canal in 1771, the small trows were used extensively to carry salt from Droitwich to the River Severn and on to Gloucester for transhipment. A special type of trow was developed for the Droitwich salt workings; commonly referred to as 'Wich barges', they were sloop rigged. Many of their deck fittings were adorned with very ornate carvings, fashioned by the boat builders, and the tiller in particular was exquisitely finished. But this type of vessel became obsolete many years ago.

Developments at Sharpness resulted in the port becoming very important as a trading centre for sea-going vessels in addition to Gloucester. Inland waterway improvements in the area included the construction of the

61 A Severn trow, *Avon*, in full sail at Bristol in 1887.

Gloucester & Berkeley Canal. Much of the traffic from the Stroudwater Canal used the canal route to Gloucester rather than the tidal Severn. The vessels that mostly used the Stroudwater Canal were known as 'Stroudwater barges', and these changed very little during the century or so that they existed. They had a rather high stem on bluff bows, were rather shallow draughted and without side-decks or catwalks and coamings. Of carvel construction, they were 70 feet long by 15 feet 6 inches wide, with a carrying capacity of 75 tons. The accommodation for the crew was provided in a cabin below the aft deck. They continued in regular service until the early 1940s, several of them remaining for a few more years in the dock area of Gloucester.

The smaller trows did not venture out into the estuaries, and were provided with an open cargo hold, without coamings, headledges or side decks, the cargo hold being protected by canvas cloths. The larger craft that did work in the Bristol Channel area, and on occasions around the coast to South Wales and Devon, were fitted with hatch-covered holds, and known as 'box'-type trows. They were all constructed of English oak and elm, as was the common practice with most sailing barges, and they worked as much as possible under sail, but inland, when the wind was not favourable, they relied first on men hauling from the banks, and later on horses. The trows, especially the larger ones on estuarial work, usually had a crew of three. The men were accommodated in a cabin under the fore deck, while the skippers sometimes occupied a small cabin aft. The vessels were not fitted with leeboards, but had a false keel unit, which was secured by chains and was lowered over the side

62 The stern section of *Water-Witch*, a Severn trow at Droitwich wharf about 1900.

when the vessel was sailing to windward – otherwise the barge would be blown off course.

In the 1860s from the River Avon section of the Kennet & Avon Canal, tar was shipped to various destinations, including across the Bristol Channel to Cardiff and Newport. A distinctive variation of a trow evolved on this traffic: using sails, it had bluff bows with a raised fore-deck and a round counter stern. The helmsman and the steering wheel were housed in a small wheelhouse aft. In the early 1920s these 'Avon Tar Boats' were converted to diesel power with the installation of early Widdop engines. They continued in use although in diminishing numbers until 1967, when they were finally withdrawn from service.

The first recorded successful attempt at using a barge made of iron was in July 1797, when a test vessel was launched on the River Severn at Coalbrookdale. It was named *The Trial*. The vessel was the result of many years' work by a Mr John Wilkinson, who had earlier tried without success to use this new building material in Lancashire, with a much smaller vessel. Eventually a trow was constructed of iron in the 1840s, but timber remained the accepted building material for trows until the end, the last one being built a little after the turn of the century.

From the late nineteenth century the widespread use of steam tugs on the

63 At the turn of the century at Worcester, showing the *M V Atalanta*, with her cargo being transhipped into narrowboats.

River Severn resulted in many trows being unrigged. From then on they relied entirely on tugs for propulsion, except for the isolated few that retained their sails. A few early attempts were made at converting trows using diesel engines, but they were not popular, and only a few were converted. The last few trows in existence, used as dumb barges, continued in service until the late 1940s. Since then the trow has all but vanished from the scene. A few of the rotting hulks remain in backwaters, although one 71 feet long with a beam of 18 feet which was used as a workshop for many years is fortunately to be restored; it is at present scuttled in Diglis Basin, Worcester.

A sailing barge used mainly on the Stover and the Hackney Canal, along with the River Teign, in South Devon was another South Western vessel. These were 56 feet long by 13 feet 6 inches, and for many years each transported some 30 tons of china clay. The 'Teignmouth keels', as they were known, had round bows and a flat stern, with a single square sail. It was a not uncommon practice to have them hauled from the banks, but in their last years of operation on the River Teign a diesel tug was used to haul them. They ceased operating in the early 1930s.

Another localised vessel was the 'Bridgewater and Taunton canal barge' which was used for trading until 1906. Measuring 53 feet by 13 feet, it was a flat-bottomed clinker-built craft, loading 22 tons at a draught of 3 feet. A

TONY
TONY
TONY
TON

64 *Tony* an early operational steel-constructed tanker on the River Severn in 1925. Powered by a Widdop 50bhp diesel engine, this motor barge was built by Richard Dunston Ltd of Thorne, Yorkshire, and was 74ft long and 14ft 2in wide.

65 The last regular traffic on the River Severn is grain from Avonmouth and Sharpness to this mill at Tewkesbury, using three motor barges and two dumb vessels. The 1970 photograph shows two of the fleet; on the fore-end of the hold of *Deerhurst* can be seen part of the automatic chain conveyor with which these vessels have recently been fitted to ease discharge problems.

simple craft, it was fitted with a small deck area both fore and aft.

Sailing barges also operated on the River Tamar, located between Devon and Cornwall from the thirteenth century, and they had their heyday during the nineteenth century. The vessels were employed mainly to transport sand for agricultural use upstream from the coast, although many diverse cargoes were carried, including manure from Plymouth. The barges had a flat bottom, with a straight stem that was slightly raked, and a transom stern, and were operated by a crew of two. They had a single mast that carried a mainsail and a foresail.

For centuries barges have worked from Barnstable and Braunton a little after high tide to the gravel banks at the Taw and Torridge estuary. While the tide ebbs and exposes the beds of aggregates, the vessels sit high and dry, waiting to be loaded. For many years the vessels were loaded by shovel, now in the days of modern handling equipment, the last few remaining vessels operating are loaded by elevator. With the tide flooding, the barges are floated

NORA EASTING
NORA EASTING

66 *Nora Easting*, seen here at Stanley Ferry on the Aire & Calder Navigation in 1975, is a modern, large capacity steel barge, built by John Harker Ltd of Knottingley for Flixborough Shipping Ltd. With a carrying capacity of 300 tons it is powered by a Gardner engine. In 1976 it was sold to an operator on the River Severn, and renamed *Chaceley*. Before commencing operations for its new owners, an automatic chain conveyor and hoppered bottom was fitted to the hold, enabling 280 tons of grain to be discharged without any trimming.

off the quickly-covered banks, to return to the wharves with their loads. The work once executed by sailing craft is now done by the three remaining vessels which are all motor barges. Of these flat-bottomed, diesel-powered craft, two are of steel, and one is of timber construction; *Result* is surely the last timber-constructed barge in daily commercial use anywhere in Britain.

From the late eighteenth century, narrowboats, known locally as 'long boats', have always traded on the River Severn, as many of the connecting waterways are narrow canals. The narrowboats worked regularly for many years into Gloucester, and up to the 1930s it was a common sight to see a diesel tug hauling a string of them up and down the river to Worcester and Stourport.

In the 1920s and 30s there were tremendous tonnage movements on the River Severn and the waterway to Gloucester and Sharpness. Many of the old wooden vessels at that time were replaced by steel dumb barges. One example of the specialised requirements that began then was the transport of oil. Oil from Avonmouth and Swansea passed through Sharpness and was delivered to storage depots in the West Midlands. Deliveries to Gloucester commenced in 1925, and further upstream on the Severn to Worcester and Stourport in 1927. New bulk oil tankers, both motor and dumb vessels, were obtained for the work. Vessels working up to Worcester were limited to a maximum length of 137 feet and width of 22 feet, working on a draught limit of 8 feet, while those working upstream to Stourport were limited to a length of 89 feet and width of 18 feet 11 inches. The peak years for the oil traffic were those between 1942 and 1955. A network of pipelines installed in the early 60s to move the oil led to the traffic on the River Severn above Gloucester terminating in 1967.

After nationalisation of the waterways in 1948, the British Waterways Board inherited a mixed fleet of vessels, mostly old steel motor barges. It was decided to replace the majority of them with large dumb barges which were to be hauled by modern diesel tugs. Eventually a serious decline in traffic over the next few years resulted in virtually the whole fleet being sold. The decline in traffic hit both the private and public sector, and many private operators ceased trading. The last traffic off the River Severn and up the Lower Avon was grain to a mill at Pershore from Gloucester. That finished in 1972, bringing to an end the commercial carrying activities of that connecting waterway after 350 years.

Not only has traffic declined on the upper Severn, but Gloucester too has lost a great deal. For many years timber was a substantial traffic into Gloucester, but since the 1960s only very occasionally does a coaster bring a load into the docks. Now on the inland waterways of the South West, on the Gloucester and Sharpness Canal and including Gloucester Docks, traffic is in the region of 300,000 tons, while further upstream, on the River Severn it is now about 25,000 tons annually. The last remaining regular traffic is grain ex-Avonmouth, which is transported in five vessels to Tewkesbury. The vessels in use include three motor barges, two of them built recently. The latest one, *Tirley*, is of welded steel construction and powered by a Gardner engine; it is 128 feet long, and was built in 1972 by John Harker Ltd on the Aire & Calder Navigation. The present owners obtained the vessel (formerly named *Thealby*) in 1976.

Vessels of the Far North

NORTH of the accepted location of most of Britain's inland waterways were a number of isolated pockets on which commercial inland craft operated for a considerable period. One of these scenes of activity was the River Tyne, on which worked craft that never went out to sea. These craft were the 'Tyne keels' which carried coal from colliery staithes, located well up the River Tyne, down to the river mouth at Shields. There the cargoes of coal were transhipped into anchored coastal vessels to be taken south, to London in particular. Coal has been shipped from the River Tyne since the middle ages. From the very beginning, the coastal vessels were unable to proceed inland to the riverside coal-loading staithes, because the river was far too shallow. Therefore the need arose for a small shallow-draughted vessel, able to work up and down the river, and the Tyne keel served this purpose.

The keels were not very sophisticated craft. They were of clinker construction, pointed at both ends, and they were usually built to a maximum length of 40 feet, with a beam of 15 feet. They could carry in the region of 20 tons. They had a small fore-deck, aft of which was the mast for the single square sail. The shallow cargo hold was without any coamings or side-decks. Below the stern deck was a small cabin for the usual crew of two. The steering was by a single oar or sweep, although in later years a tiller and rudder were fitted. For the 15-mile journey it was usual to wait for the tide and the resulting current to assist. These vessels remained a common sight on the River Tyne for five hundred years, until they stopped regular working a little before the First World War, although some craft existed well on until the early 1940s.

Further north, the Scottish canals consist of the now closed Forth & Clyde, and Union Canals and the open Crinan and Caledonian Canals. The Caledonian Canal is now the only canal linking the east and west coasts of Scotland; of its 60½ mile length, 38 miles consist of freshwater lochs. On its route, at the western end, are the biggest staircase locks in Britain, lifting the canal 64 feet. This canal was opened in 1822, and from the beginning relied upon the passage of sea-going fishing vessels and passenger boats. The steamboat services for many years provided lucrative traffic, and now fishing vessels continue to use the waterway, as do an ever increasing number of sea-going pleasure craft.

The Crinan Canal, which is only 9 miles long, cuts across the Kintyre Peninsula in Argyll on the west coast, connecting Ardrishaig on Loch Fyne and Crinan on the Sound of Jura. The waterway provides a short cut, the alternative being a lengthy passage round the Isle of Arran and the Mull of

67 This horse-drawn cargo vessel, a 'scow' of timber construction, is at the Almond Aqueduct on the Union Canal in Scotland, which closed in 1933. The vessel has the smallest of hawse timbers, very few timber heads and only a stove in the aftcabin, the forecabin being used for storage purposes only. The horses in the foreground have no connection with the waterways – they were ploughing the field.

Kintyre. The canal was opened in 1801, and for many years was busy with the passage of fishing vessels, coasters and passenger steamers. Now the canal relies increasingly on the passage of sea-going pleasure craft and a decreasing number of fishing vessels.

For many years the Crinan Canal was used by some of the most interesting of the Scottish-built vessels, the 'puffers'. These were small steam-powered craft, very similar in shape to the former Weaver 'packets', but having more free-board and a higher superstructure. The puffer is now almost extinct, but a few examples remain, lovingly cared for by enthusiasts. The first puffer, a stern wheeler, was introduced in the early 1830s, and from then on they continued to increase in number. Their two-cylinder compound engine, exhausting directly up the funnel, produced the familiar puffing sound from which they derived their name. Stern wheelers died out early, being replaced by the screw-driven propeller puffers.

The puffers worked from the River Clyde to the west coast islands, serving the needs of the islanders, and bringing them every possible type of merchandise, from fuel to livestock. The vessels did not have a flat bottom, but they had the ability to be beached in a sheltered bay, enabling the cargo for the islanders to be transhipped. They stayed on the sandy shore until the next high

68 This photograph shows Carron Company powered 'scow' No 16 passing the lifting Camelon Bridge. On the right is *Afghan*, another steam powered vessel, owned by Gillespie of Camelon, who also owned the warehouse and wharf. Both are wooden-constructed vessels; only *Afghan* has hold covers, while No 16 has a very high headledge at the fore-end of the cargo hold.

69 This wooden-constructed 'scow' was being used for general maintenance duties on the Union Canal in 1951.

water, and then the puffer would go on to the next destination. The puffers went on using the Crinan Canal until the early 1970s. They had a simple layout, with engine and wheelhouse positioned aft, behind the tall funnel. Next came the deep cargo hold, which could carry up to 120 tons, and from which the cargo could, if the need arose, be discharged with the aid of the vessel's own derrick, utilising the vessel's own mast and jib. The crew, usually of two or sometimes three men and a boy, were accommodated in a cabin in the forecastle, which was usually referred to as the den.

Puffers were not unknown on the eastern seaboard of Scotland, having gained access to the area via the Forth & Clyde Canal. The vessels that made that journey were usually called 'gabbarts', and they were a little smaller than the ordinary puffers, because of the lock limitations; they were only 66 feet long by 18 feet wide, with approximately 6 feet draught. Their passing is very much regretted, for they were vessels with great character. 'Gabbart' is also a term that referred to the river lighter used on the River Clyde, and timber-constructed, sloop-rigged craft that operated on the west coast of Scotland.

The everyday cargo-carrying dumb craft of central Scotland were called 'scows', 60 feet long with a beam of between 12 and 13 feet 6 inches, carrying up to a maximum of 80 tons. The craft were usually double ended, with small deck areas at both ends, and most of them were fitted with hatch covers over the hold. Originally of timber construction, from the early nineteenth century they were built of iron. The last one which was built, of steel, was for the Forth & Clyde Canal, and entered service in 1948.

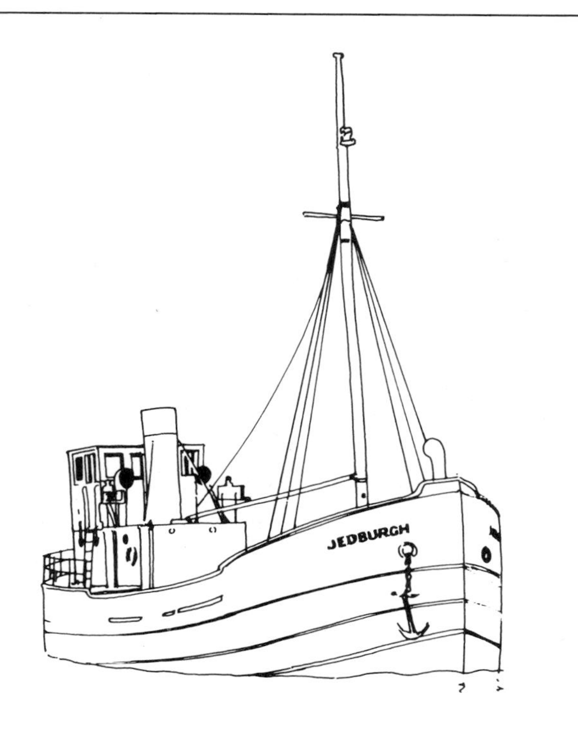

Scottish canals 'puffer'

Vessels of the South East

THE INLAND WATERWAYS of the South East are mainly based on the River Thames. This river, in particular, has been used ever since the movement of goods by water became an accepted mode of transport. The oldest known Thames craft is the 'peter' boat, a small clinker-built double-ended rowing or sailing boat that was developed solely for fishing on the river. The boats varied in length, up to a maximum of 25 feet, and are mentioned here, because, besides being early inland waterway craft, they possessed a small hold amidships for the 'cargo'. It is assumed that the small fishing vessel got its name from St Peter, the patron saint of fishermen.

Throughout history, because the River Thames ran through the capital of the nation, it has been a vital channel for commerce, and has been used increasingly by sea-going vessels which have come to the great port with their cargoes over the years. The foreign trading vessels gradually got larger, and the risk of their being stranded on the mud became greater, for they found it increasingly difficult to get to the riverside wharves. This resulted in a reduction in the number of vessels proceeding upstream from the pool below London Bridge. Here it became the standard practice to unload the ships in mid-stream, transferring the cargoes to small river vessels which took them either to local riverside wharves or to destinations further upstream.

Originally the cargoes destined for local delivery were transported in small vessels, manned by one or two men, who rowed or punted to their destinations with the assistance of the tide and currents. Later the cargoes that were consigned to towns and villages a little further afield, including locations on the tributaries, were transhipped into larger craft. These longer distance cargo-carrying vessels which regularly worked to the west of London, and up the river beyond the tidal reaches as far as Oxford, were known as 'West Country barges', and in time became the traditional cargo-carrying craft of the river.

Upstream of London in places the river was exceptionally shallow and this limited the draught of the vessels considerably. In 1768 an Act of Parliament limited the draught of vessels using the upper Thames to 3 feet, but a few years later this was changed to 3 feet 10 inches. The 'West Country barges' were very much like large punts, flat bottomed with swim-headed bows and a very large rudder. Because for many years there were no locks on the upper reaches, the only restrictions to the size of the vessels on the river were natural ones. Depending therefore upon the requirements imposed by the usual operating area, the vessels were built with many different cargo capacities, varying between 25 and 200 tons.

On the wide river sections a single square sail was used whenever possible in preference to hauling from the bank. On the upper sections of the river the vessels relied upon men to haul the vessels along until the early years of the nineteenth century, when it became the general practice to use horses. Because of the large size of some of the shallow draughted vessels, and the considerable number of shallows encountered, the gangs of men doing the hauling sometimes totalled between 50 and 60, but when horses were used, a maximum of about a dozen were needed.

For a considerable period of time the trade on the river continued in very much the same manner, with gradually more distinctive classes of vessels evolving. One type was the 'Thames lighter', used on localised transport work; the crews had the ability to control their rather clumsy swim-headed vessels in the worst of conditions. Such craft were devoid of any graceful lines or any beautiful decorations, and they had not even a cabin.

Another type of vessel was the sailing barge, very similar in construction to the lighter, with flat square ends that raked outwards to overhang the water. A small fore-deck was fitted with a winch for hauling up the mast or anchor, and there was a small living cabin for the crew below the deck aft. The mast was usually located well forward and stepped in a tabernacle, so that it could be lowered to pass under bridges. At first rigging consisted of a square sail, but this could only be used when the vessel was going in the direction of the prevailing wind.

Over the years more and more sailing barges worked on the estuary to the Medway ports, and these were generally of a larger size than those that went upstream to the west of London. The use of leeboards became common on the craft that worked on the estuary. Working to the Medway ports, a considerable number of craft were employed transporting chalk, which was obtained from quarries in Kent and taken to Gravesend. These vessels were known as 'chalk barges', and had a carrying capacity of about 40 tons. Other vessels, between 45 and 55 feet in length, transported agricultural products, with hay and straw bales stacked sometimes as high as 12 feet above deck level. These vessels were used extensively during the first half of the nineteenth century. They were shallow draughted and this made it possible for them to venture up riverside creeks, to load direct from adjacent fields. Some of these craft were still in use after the turn of the century.

The passing years brought further improvements in vessel design, which resulted in craft being constructed with curved sides, the fore and aft ends being narrowed, and, increasingly, having hatch covers fitted. Spritsail rigging became common; with this the mast was placed one-third of the distance from the bow and the sail had additional support from a diagonal pole, the base of which was supported from the mast, so that the pole was at an angle of about 60°. This pole extended the canvas sail to a peak, higher than the top mast connection. With this rigging an additional foresail was used as standard, giving a system which could conveniently be managed by one man if the need arose. A vessel with round bows was increasingly coming into use, although the old design, with swim-headed hull, continued to be built.

In the mid-nineteenth century the sailing barge, as it is known now, evolved. Of most charming lines, the vessel had a vertical stem post and a transomed stern, and was approximately 82 feet long with an 18 foot beam.

70 The sailing barge *Webster* is moored at Barnes – a topsail barge with topmast lowered. Although she is wheel-steered, a small mizzen, used to assist tiller steering, is positioned on the rudder head. The vessel was built at Lambeth in 1863, able to carry 42 tons, and was owned by Wakeley Bros of Southwark, who were engaged in transporting rubbish.

The mast was usually about 30 feet high, but an additional top mast for a topsail rig would add approximately a further 22 feet. A few vessels, first of iron and afterwards of steel, were constructed and in later years some had a diesel engine installed. For so many years steering of the Thames sailing barge had been with a tiller, but during the latter period of the vessel's commercial operations it was with a wheel.

The Thames sailing barge was to become one of the most decorative of vessels. It was made of the best materials, including Oregan pine and English oak and elm. Resplendent under its red brown canvas, it was used throughout the Thames estuary and along the east coast to many rivers and creeks, as well as to small ports in Sussex and Hampshire. Since the last war, it has been one of the most used of the sailing barges, one of the regular long distance runs from London being the transport of grain to Peterborough on the River Nene, which continued until 1966.

The barges continued in commercial operation until 1971, when *Cambria*, the last one that worked regularly under sail, was withdrawn. Approximately fifty Thames sailing barges remain in existence, the most of any type of sailing barge. They are now used for various purposes, from houseboats to providing cruising holidays under sail. Much has been written about these craft, far more than any other estuarial sailing barge, and several preservation groups exist which continue to research into the history of the vessels.

An inland waterway in the South East that had vessels of particular interest

71 This photograph of the yard of Richard & Edward Talbot, barge builders, at Caversham Bridge was taken pre-1869, for in that year the stone bridge was replaced by one of wrought and cast iron. The vessel with all the men on board is no doubt in the final stages of completion; note the rudder which is rather large and of crude construction.

72 *Phoenician*, a Thames sailing barge built by Wills & Packburn in 1922, was winner of the Thames match in 1932/3/5/6 and Medway 1931/2/5/6. This vessel traded commercially as a motor barge until 1973, when she was re-rigged.

was the Stour (Essex) Navigation. The vessels that worked on that waterway, usually referred to as 'Stour lighters', have been immortalised in the paintings of John Constable, who featured the craft prominently in a number of his famous works. The paintings are all interesting, but one worthy of special note is 'Boatbuilding near Flatford Mill', in which the artist captures a rare scene showing a small yard with a vessel under construction. Here the skills and expertise of the boatbuilder are preserved for posterity. In fact the majority of the Stour lighters were actually built at the yard at Flatford. The vessels measured approximately 47 feet in length with a beam of 10 feet 9 inches, and a draught of 2 feet 6 inches when fully loaded. They were usually towed in pairs by one horse, but, as several paintings show, occasionally a small square sail was used. The last lighter to be used carried a cargo to Dedham in 1928. Recently one craft that had been sunk for many years was lifted, and it is now to be preserved.

To the west the Wey Navigation also had a distinctive type of vessel, a derivation of the former sailing barges of the upper Thames. Until the very end these craft worked regularly from the Wey, to the River Thames and the London docks. The Wey vessels, of carvel construction, had bluff bows and a flat bottom, and were capable of transporting 80 tons on the river, but were limited to 50 tons on the Wey Navigation. They were strongly built, with a small 'D'-shaped transomed stern. Similar craft also worked on the adjoining waterway, the Basingstoke Canal. These were of wooden construction and fitted with hatch covers. The last few Basingstoke Canal vessels continued in use until the early 1930s; they worked to a maximum draught of 4 feet 9 inches, and were a total length of 73 feet 6 inches with a beam of 13 feet 10 inches.

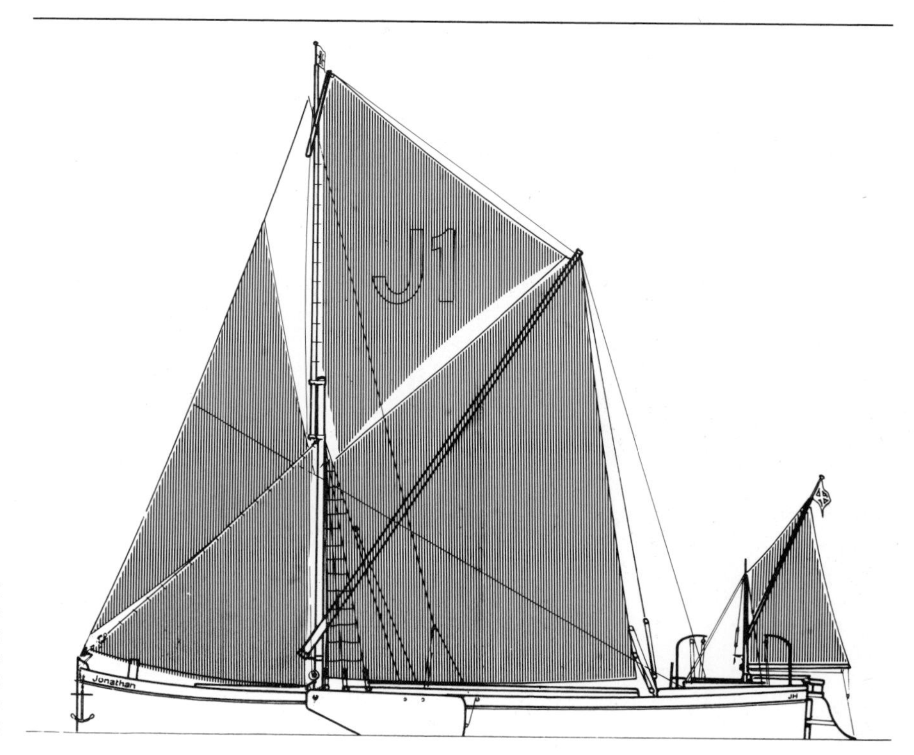

Rigging of a Thames sailing barge: the example shows a small river staysail barge (without a bowsprit). The sail areas are as follows:

Mainsail	1130 sq ft
Foresail	310 sq ft
Topsail	600 sq ft
Mizzen	130 sq ft
Staysail	460 sq ft
Total	2630 sq ft

On the Wey Navigation, traffic finally came to an end in the early summer of 1969, when the last three vessels in use were withdrawn from service. They had been employed transporting grain from the Port of London to Cox's Lock Mill at Weybridge. As they were dumb vessels, tugs towed them on the river to Thames Lock, Weybridge and from there they were hauled by horses the rest of the way to the mill.

On the upper reaches of the River Thames were a number of connecting waterways, including the Thames & Severn, the Wilts and Berks Canals, as well as the Kennet & Avon Canal. This assured many varied workings and kept the craft operational. One of the most unusual was coal from the Somerset Coal Canal that was regularly transported from London by narrowboat, via the Kennet & Avon Canal.

From the mid-1930s the Grand Union Canal was provided with wide-beamed locks all the way from London to Birmingham. The Grand Union Canal Company, aware that wide-beamed craft would be more suitable than the narrowboats that had always been used on the waterway, were very hopeful for their first motor barge when it was put into operation. This motor barge, *Progress*, was only the first of a large new fleet that was to be built. Unfortunately, despite the wide locks, the greater part of the channel on the route to Birmingham was not wide enough for the vessel when loaded, and so it was not a success in operation. The *Progress*, with a wooden hull, had wheel steering and was fitted with a Junkers diesel engine. It was relegated to

73 On the Wey, a William Stevens & Sons transom stern barge is under way, hauled by a small tug. Usually the barges were hauled by horses. The small vessel also on tow is *Fly*, the waterway's maintenance vessel in later years.

74 At Regents Dock in London in the 1930s, a number of 'pairs' of Fellows, Morton & Clayton Ltd boats are loading or waiting their turn. The cratch section at the fore-end of the hold, along with towing masts and stands for the top plank, can be seen. A few lighters, the simple swim-headed dumb vessels, share the scene.

maintenance work, until sold to a private individual in the 1960s. The Grand Union Company therefore had to rely on narrowboats for all their long distance work. On the Thames these were commonly referred to as 'monkey boats'.

The lowest section of the Grand Union Canal could cope with wide-beamed vessels, and for a considerable length of time many dumb vessels worked most satisfactorily. These continued in operation until well after the last war, although their area of operation has been reduced gradually until now only a very short distance is used.

The River Lee, the first waterway in Britain to receive an Act for navigation improvement in 1424, is a successful commercial waterway today. Originally it relied on small sailing vessels; these were followed by lighters 70 feet in length, with a 12 foot beam, able to carry 45 tons. In the early 1930s lock improvements were completed, and this allowed a larger capacity vessel to operate, so a considerable number of wooden vessels, with 3 inch planks on iron frames, were built, 88 feet in length and with a beam of 15 feet. At the

75 These swim-headed lighters, the universal dock vessel, are located at Clement Tough's yard at Brentford. Note the three tugs, and in particular the draught requirements of the small one.

same time, for operation on the connecting waterway, the Stort, a number of smaller steel dumb vessels were built.

The upper Thames was for a considerable time one of the busiest of rivers for the movement of cargo, but things have now changed. In recent years the trade of the Port of London has declined, with the closure of many docks since 1960. This included the Regent's Canal Dock, which provided much traffic for narrowboats. Also many of the wharves and warehouses on the river have been closed. All these closures had a considerable effect on the trade on the waterways in the area. But the Lee Navigation, from the Thames at Blackwall and Limehouse, allows a through route to Enfield, and provides a transport medium that avoids the most congested area of London. It is operated by lighters of up to 140 tons capacity, carrying both import and export cargoes.

Also the Grand Union Canal, although now nominally a cruising waterway, connects with the Thames at Brentford, to the west of London. Here again lighters carrying up to 140 tons transport cargoes between ships and docks, including Tilbury, to the British Waterways depot a short distance along the canal. Another waterway still in use is the Deptford and Crayford Navigation which now carries in the region of 100,000 tons annually in similar lighters.

The modern lighter, therefore, carries a vast amount of the remaining traffic. It is built of steel, and the basic design has remained unchanged for a very long time, with a flat bottom and a square chine at the bilge, along with straight overhanging bows (i.e. swim-headed), and no cabin. On the Thames very little commercial carrying is now done in small vessels, compared with previous workings, so that the number of lighters and tugs has been reduced dramatically. However a number of companies do own large fleets, one example being Humphery & Grey, who now operate a fleet of 120 steel lighters and five tugs on 60 miles of the river, between Sheerness in Kent and Brentford in Middlesex. In addition there are a number of tank barges, the majority of which are used on bunkering duties. One company, Allantone Supplies Ltd, ordered a new 300-tonne capacity vessel from John Harker (Shipyards) Ltd of Knottingley, Yorkshire, in 1978.

Vessels of East Anglia

For several hundred years, from the sixteenth century, the trading vessel most commonly seen on the waterways of East Anglia was the square rigged 'Norfolk keel'. These vessels, of clinker construction, were of two basic hull designs, either pointed at both ends or with a small transom stern. With a small cabin in the fore-end, the cargo space was capable of holding between 30 and 40 tons, and was without any shutts or hatch covers, while the mast was positioned in a tabernacle. They were of robust construction, around 50 to 55 feet in length, with a beam of about 14 feet. Gradually throughout the nineteenth century their numbers were depleted until by the early 1890s they had become extinct. The Norfolk keel was replaced by the more manoeuvrable 'Norfolk wherry' whose sails were to dominate the skyline for nearly a century.

The wherries ruled supreme on the Eastern waterways of the region, with

Norfolk wherry

76 Gangs of Fenland lighters moored on the River Little Ouse at Brandon Bridge.

practically every industry relying upon them. Farmers in East Norfolk had beet, grain and potatoes shipped, while other cargoes ranged from building materials to coal. At the turn of the century well over 300 wherries operated on the rivers Ant, Burem, Thurne, Waveney and Yare, trading to Norwich and the towns and villages throughout the area, connecting them to the sea and export facilities through Yarmouth. The wherries were mostly crewed by a man and boy, or a husband and wife team, who knew every inch of the waterways. Like the majority of inland sailors they were an independent breed, happy and content, despite the long hours, coping with wind and tide. In spite of their bulky size, the vessels were some of the handiest of sailing vessels on the inland waterways, able to sail very close to the wind.

The rigging was very simple, consisting of a single gaff mainsail, unstayed, except for the fore-end, that had a single halyard and a hand-operated winch. The mast, usually of pitch pine, could weigh up to a ton. It was counter-balanced with weights that could weigh as much as $1\frac{1}{2}$ tons. The large brown or black single sails were dressed with a mixture of fish oil and coal tar, which preserved them and made them last for a considerable time. The size of the clinker built wherries varied considerably, for they could carry between 12 and 83 tons. The smaller ones were usually in the region of 30 feet in length, while the larger ones were usually about 60 feet long with a 17 foot beam. The largest recorded wherry ever built was *Wonder* of Norwich, launched in 1878. It measured 65 feet in length with a 19 foot beam, having a draught of 7 feet when loaded with 83 tons of cargo.

The hulls of the wherries were tarred black; they were rather pointed at each end, and well rounded with a good flair at both the bow and stern. They were usually vee bottomed, and on the whole worked at a rather shallow draught. At the stern was a massive wooden rudder and tiller. The wooden fixtures and fittings were painted red, white and blue – a colour scheme which was most

77 Three separate gangs of Fenland lighters in Alexandra Dock, Kings Lynn.

78 At Cambridge wharf about 1900: on the right the Eastern Counties Navigation Company's wooden-constructed steam barge *Nancy* is being unloaded. Behind is a narrowboat steam tug and a Fenland lighter loaded with barrels.

popular. Spanning the hold, aft of the mast, was a large cross member on which was carved the vessel's registered tonnage and number. At the bows, the wherries nearly always had a traditional splash of white, to form a quandrant or eyes, which helped with identification of vessels during the hours of darkness. The last few wherries in commercial use ended their trading life with the seasonal transportation of sugar beet in the late 1950s.

For many years before the coming of the railways, the Fens area of East Anglia relied extensively on water transport for both passenger and freight movement. The packet boats operated on the rivers to centres of habitation such as Bedford, Bury St Edmunds, Cambridge, Ely, Huntingdon, Kings Lynn and Northampton. Cargoes for the most part were transported in the shallow

79 Tank barges *Enid* and *Lizzie* load gas-water at Cambridge gasworks in 1910. The steam tug *Olga* waits alongside ready to haul these dumb vessels to a fertiliser factory at Kings Lynn.

'Fen lighters', that were 42 feet long with a 9 to 10 foot beam at the bottom, getting broader by about 1 foot at the top. The Fen lighters generally worked in a 'gang' of four or five, and on occasions with a sixth, a 'Butt' boat. This Butt boat, approximately 30 feet long with swim ends, always had space left on board the vessel to ferry the horse from one side of the waterway to the other. When underway, the towing line, which was between 150 and 250 feet in length, was secured to a strong hauling mast about 12 feet high, positioned a little forward of amidships in the leading vessel.

The second vessel was secured as tightly as possible to the leading one, with the bows hard up against the stern. A thick strong rope, tied in a figure of eight fashion, was used. The second lighter was the 'house lighter', an ordinary lighter, but provided with a small cabin to accommodate the crew. The second lighter always had a strong heavy pole, about 30 feet long, protruding over the bows and half way along the leading vessel, to just short of the towing mast. This pole, being the steering pole, was provided with rope and was securely lashed – unless the course of the train of lighters required changing. Over the bows of each of the following lighters protruded a shorter pole, extending only about 6 feet. The third lighter, the 'Hollip', was an ordinary lighter but occasionally was fitted with a small cabin, called a hutch, under the bow staging. Here the horse boy slept. The fourth vessel was a conventional Fen lighter, without any special purpose other than to carry cargo.

In the 1890s on the River Nene, between 50 and 60 lighters made deliveries every week at Peterborough. The lighters were a very practical and suitable method of transporting cargoes, and they operated for many years. Change came when the steam tugs started to replace horses. They speeded up their operation considerably, and the steering system and accommodation became obsolete as the tugs were able to haul a number of boats in a manageable train. A few steel lighters were introduced to replace some of the wooden lighters, while others were withdrawn. The River Nene continued as a commercial waterway, and new traffic was encouraged. Conventional steel motor barges made their appearance in the 1930s. Grain in large quantities continued to be transported between Northampton and Wellingborough, and this required the use of more and more narrowboats. These had appeared on the waterway after the opening of the narrow canals and they continued working on this traffic until the last load carried in 1967. The final commercial carrying operation on the River Nene, using steel motor barges, was actually the movement of stone, mostly for river bank protection, but that terminated in the early 1970s.

Another river in the area that experienced great changes in the type of vessels operating was the River Yare. In the 1890s a considerable number of steel dumb barges with pointed stem and stern were built for operation on this river, replacing many of the former wooden craft. They were of various sizes, for different operators, ranging from 40 to 75 tons carrying capacity. The steam tugs, which came onto the scene at the same time to haul the dumb vessels, were typical of many built at that time; they were around 55 feet in length, with tall funnels, and the hulls were of pleasing lines, with fine shear and low round counter sterns. They were used mostly for work between Yarmouth and Norwich, the last working only coming to an end in the mid 1960s when coal ceased to be carried. Trade still continues on the river to the developing port of Norwich – 500 ton motor coasters bring cargoes of animal feeding stuffs, fertilisers and timber, with occasional loads of scrap metal and barley for export. However, all workings by small inland waterway commercial carrying vessels in the area have now ceased.

Narrowboats

THE earliest canal boats were the 'starvationers', simple double-ended craft that carried coal in the Duke of Bridgewater's flooded mines. These mines were at Worsley near Manchester, opened in 1759, and extending at different levels for nearly 46 miles. The vessels were very rudimentary in construction, the longest some 50 feet in length with a 7 foot beam. Their planking was attached to the frame and ribs in such a way that the latter could be clearly seen, giving them a lean and 'starved' look, whence they got their name.

The first navigations had locks 14 feet wide, but when Brindley engineered the first of the inland canals, he built locks only 7 feet wide to accommodate craft that no doubt were only a little more sophisticated than the starvationers. What the exact reasons were for determining the size of locks at about 71 feet long and 7 feet 2 inches wide it is impossible to say, but as the locks were only half as wide as the river locks, they would in turn use only half as much water. This was important on the canals, for practically every drop of water had to be provided from artificial reservoirs. Also as the first canals extended across the country to link up the main estuaries of the Thames, Severn, Mersey and Humber, many tunnels had to be driven through hills, and while the early engineers were capable of constructing a tunnel 7 feet wide, twice the width was no doubt an impossibility. It was not until many years later that construction of wider tunnels became common.

80 Nearly at the end of an era: *Snowflake*, owned by Potter & Son (Runcorn) Ltd, in Manchester in the 1950s. A family boat, mother stands in the open doorway of her home, while her son, who was not to carry on the traditions of the past, sits and stares. The traditionally decorated cabin, with roses and castles and all the fittings, is a first-class example of a working butty.

At first the boats began work on the partly completed canals, and so the journeys were relatively short and there was no need for living accommodation. But when the network of canals was extended, a need arose for boats with temporary living quarters, and this resulted in the development of the first of the narrowboats as we know them. Vessels capable of carrying a load of about 30 tons were built some 70 feet long with a 7 foot beam, at a loaded draught of 3 feet 6 inches, the exact dimensions varying according to the locks of the canal on which they usually operated.

During the first century of the canal age (1750–1850) the boats of the canals were worked by men and boys. The introduction of the railways and the spread of their network resulted in much competition for traffic, so more economical ways of working the boats were sought, and many boatmen gave up their homes and brought their families to live on the boats. This saved the expense both of a paid hand, and of a house, for the family thereafter assisted with the running of the boat. Thus without the cost of a house and mate, the majority of the narrowboat skippers managed to exist in a period of low rates.

Naturally when the family lived on board, the small cabin space, only 10 feet in length, had to be fully utilised. So efficient was the resulting design of the cabin, that it remained practically unchanged for over a hundred years. At the stern of the boat was a cockpit that gave access to the small, but very cosy cabin. Entry into the cabin was through two small doors of wood graining which were gaily decorated in the traditional manner with castle scenes. They opened outwards to show the footboards on which the steerer stood. In the roof of the cabin, above the footboards, there was a sliding panel, called a slide hatch. The footboards were positioned several feet above the cabin floor, and to bridge the gap between the footboards and the cabin floor was a coalbox, used also as a step.

Along the right-hand side of the cabin was a bench with drawers and lockers fitted into it, the top of which was used as seats during the day and a bed for the children at night. Above the bench was a single porthole in the cabin side. On the left, positioned on a low stand, was the black-leaded coal range with an oven, and with the chimney going up through the roof. A slim floor-to-ceiling cupboard, built at a slight angle, was next on the left. This had a round-topped cupboard door, hinged at the bottom, so that it opened downwards to make a table.

Beyond this fixture was another floor-to-ceiling fitting, but much wider. Here was the double bed, with bedding concealed by a cupboard door, which also hinged downward to bridge the central aisle of the cabin. The cross bed, 3 feet 6 inches wide, held the mattress that was unrolled across it when required. Lace-edged curtains concealed this bed section of the cabin at night.

The interior of the cabin was decorated in grained wood finish, along with panels of roses and castles. Draped white lace curtains and coloured, china, lace-edged plates hanging by bright silk ribbons contrasted with the woodwork. The warm cosy cabin had a final touch: polished brass, which gleamed by the light of the flickering paraffin lamp.

Hundreds of narrowboats have been built with slight variations on the basic design. They were built either in large yards or in small premises where only a few men were employed, but every yard produced vessels with characteristics and shapes peculiarly their own, with just that little difference from all others.

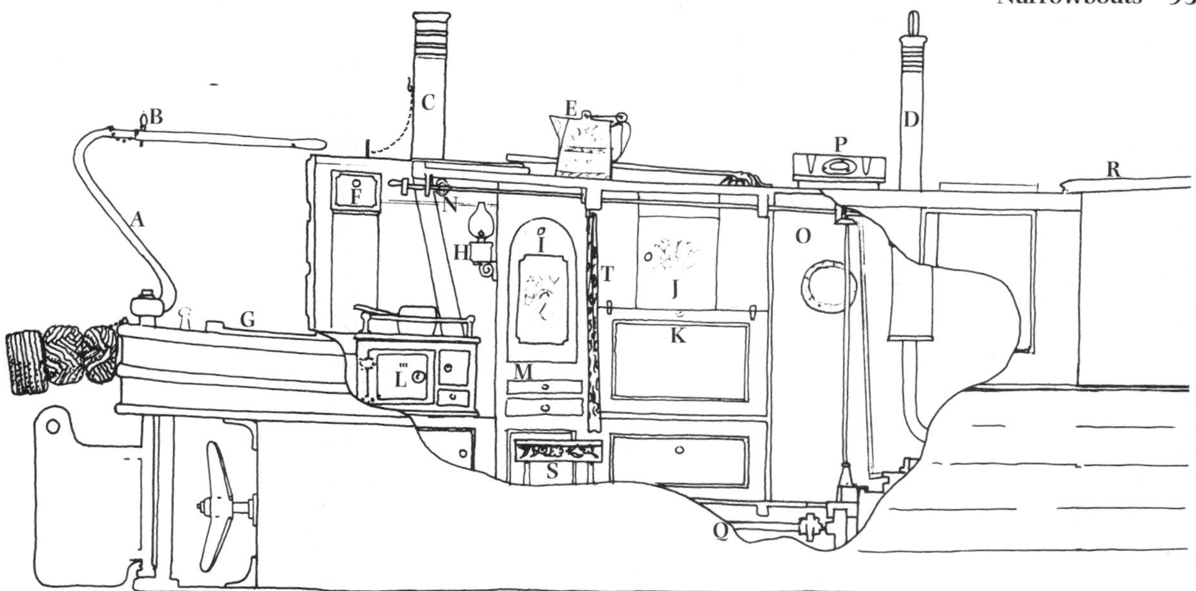

Sectional drawing through the cabin of a motor narrowboat

Key:
A Tiller
B Tiller extension pin
C Cabin stove chimney
D Engine exhaust
E Buckby can (drinking water)
F Ticket drawer (tolls etc)
G Counter
H Paraffin lamp
I Table cupboard
J Cupboard
K Cross-bed (folded up during the day)
L Cabin stove with oven
M Knife drawer
N Engine controls
O Engine room
P Pigeon box (engine room ventilator)
Q Propeller shaft
R Plank over hold
S Cabin stool
T Curtains to separate cabin at night

Construction methods, however, were similar throughout the country. The bottom of the narrowboat was usually of 3-inch-thick elm, with a stem and stern post of oak. These two were usually made and shaped from a tree bough, the workman following its natural curve rather than cutting across the grain. They were fixed to an oak keelson, which ran from the fore-end to the stern and was made up of several sections scarfed together (i.e. joined by means of a special overlapping joint). Planks that had been steamed to make them pliable made the hull shape, and each of these oak planks was designed so that it could be removed at a later date, without disturbing the other planks.

Inside the hold, connecting and holding the sides and bottom together were 'knees', strong ribs of oak, which in later years were sometimes made of iron or steel. Inside the hull was an inner planked lining with a gap between it and the exterior planking, filled with chalico (a mixture of tar, cow hair and horse dung). To prevent the sides of the hull from spreading outwards, either a rigged timber support or tensioned chains were positioned across the top of the hold. To protect the bow from knocks and give added strength, an iron strip was fitted down the outside of the stem, with steel strips fitted around the exterior to act as rubbing bands.

At the front of the boat, from the stem to the cargo hold bulkhead, it was decked over, and the area in between was used as storage for ropes, fenders, etc. Aft of this small deck, above the bulkhead, was a vertical triangular board, called the 'cratch'. In the fore-section of the hold was the towing mast, and evenly spaced back to the fore bulkhead of the cabin were two posts, known as the stands. From the cratch to the cabin top over the full length of the cargo hold was a plank, sitting on the top of the cratch and the stands. This was provided to hold the canvas covers that draped down both sides to cover the cargo hold, and which were usually secured with thin rope.

Originally all the narrowboats were constructed of wood, but later narrowboats were built with iron and then steel sides, with wooden bottoms,

FELLOWS. MORTON
& CLAYTON, LTD.
Nº 242
11316
17319
NORTHWICH

81 A number of narrowboats have been restored for museums, most of them as floating exhibits. At the Waterways Museum at Stoke Bruerne can be seen this vessel in a weighing machine. The former horse boat *Northwich* worked for many years for the large operator Fellows, Morton & Clayton Ltd.

usually of elm; these are known as 'composites'. A number of vessels were built throughout of steel, including the cabins, but usually the cabins were made of wood. Examples of all these variations still exist today.

Aft of the cabin was a small well, the cockpit used by the steerer, and fitted with a locker in the aft section for various boat stores. The horse-drawn narrowboats had what can best be described as a pointed stern, on which was hung a large wooden rudder. The large member of the rudder, known as the ram's head, was usually decorated in a geometric design, with intricate, exquisite ropework adornment. The elegant tiller of wood curved down to the sliding hatch of the cabin when in use. When not required it was reversed in its socket so that it curved upwards and out of the way of the cabin doors.

For many years the exterior of the boats have been decorated in a similar manner throughout the country. The cabin sides are painted in bright colours, with a large panel on each side showing the company's name and base, and a small inset panel bearing the numbers which indicate the boat's registration details. The lettering follows a common style, usually with shading being an important factor. Alongside is another, smaller panel, painted with a landscape scene incorporating a traditional castle and lake. The cabin roof is woodgrained overall except when a central aisle of brightly painted diamonds is incorporated.

Positioned at the rear left-hand side, the black cabin stove chimney is nearly always adorned with three brass rings near the top. Fitted to the chimney side is a brass safety chain, to prevent the chimney (which is detachable) from being knocked overboard accidentally by a low bridge or the branch of a tree. On the roof, next to the chimney, is the home of the water can, known as a buckby can, as well as the dipper, used as a general handbowl. Both are painted with the traditional rose decoration and perhaps the name of the boat, or the skipper's and his wife's first names. Another loose item usually present is a mop, the handle painted in a striped barber's pole fashion.

The remaining parts of the boat, except the lower section of the hull, are painted with either geometric designs, or moons, or playing cards, or a mixture of all three. These parts include the towing mast, stands, aft bulkhead and bows, while the cratch at the fore-end looks most attractive when adorned with the full glory of the traditional rose painting.

For many years the narrowboats relied upon the horse hauling from the towpath, and most boats worked individually. In the last quarter of the nineteenth century a number of boats were built with steam engines installed. One of the first steam-powered narrowboats was *Dart*, built for the Grand Junction Canal Company. Many more were to follow, especially on that particular waterway, usually powered by a vertical compound steam engine driving a 2 foot $10\frac{1}{2}$ inch propeller at 280 rpm.

Thirty-two steamers were built between 1886 and 1923 for the Birmingham-based Fellows, Morton & Clayton Company. They were mostly employed on a non-stop service from Fazeley Street, Birmingham, to City Road Basin, London, each vessel usually manned by a crew of four, and taking approximately 54 hours for the journey. They were 72 feet long with a 7 foot beam, the majority had an iron hull with a $2\frac{1}{2}$ inch thick elm bottom. Later, narrow beamed tugs were used extensively in some areas, especially on waterways with few locks.

82 Two narrowboats pass on the Rochdale Canal. When passing it was the usual practice for the boat on the outside to allow the tow line to sink to the bottom of the canal so that the other vessel could float over it. This photograph shows the vessels passing the wrong way, keeping to the left instead of the right. If one had been loaded it would be understandable, requiring the channel.

The great failing of the steam cargo-carrying vessels was that the power units were bulky, and, together with the fuel, either coal or coke, took up valuable cargo space, even as much as a third. They also required the constant attention of a trained boiler man or engineer, so they were usually operated with a crew of four men. In fact the steam boats were the last narrowboats that had regular all male crews.

Eventually a power unit was developed that solved most problems. This was a diesel engine, built by a Swedish company called Bolinder. It took up little space, and once started it would run for many hours without the slightest attention. The first vessel fitted with this new propulsion unit was Fellows, Morton & Clayton's *Linda*, which went into operational service in 1912. Soon afterwards the company's steamers were duly converted to diesel or they were

83 *Vulcan*, a steamer of Fellows, Morton & Clayton Ltd, is powered by a vertical compound steam engine, driving a propeller with a diameter of 2ft 10½in. The company had 32 steamers in all built between 1886 and 1923.

sold, so that by 1927 all their steamers had gone. Increasingly, from the early 1930s, four-cylinder diesel engines were fitted into narrowboats.

By the mid 1920s the design of the motor narrowboat as it is now evolved. The engine was housed in a longer cabin, forward of the living quarters, and access to the engine 'ole' was gained via two side doors, one on each side of the cabin, which could be reached by a catwalk on the gunnels, and connected with the stern deck. Because of the existence of the propeller shaft, the headroom in the motorboat cabin was less than that in the horsedrawn boat. The bows of the motorboat remained the same as other narrowboats. The aft section of the vessel when empty sat with the stern well down, to keep the propeller deep in the water. It had a semicircular stern deck, overhanging the propeller, acting as an anticavitation plate. The rudder was out of sight below

84 A single-cylinder Bolinder engine from a narrowboat is being worked on by a fitter.

the water, with a tiller of round steel, shaped like a 'Z', usually with a detachable brass extension piece at the top. The motorboat, with a shorter hold, was able to carry approximately 25 tons of cargo.

It soon became the standard practice for a motorboat to haul a former horse boat, and together they became known as a 'pair', or 'Joshers', the horse boat becoming known as a 'butty' (a buddy). Two boats meant a proportionate increase in cabin space, the butty cabin providing additional sleeping accommodation when one family operated a pair. The woman usually steered the butty, endeavouring to do her household duties in between times.

Throughout the history of inland waterways, narrowboats have worked to all parts of the system, not only on narrow canals but also on adjoining wide-beamed waterways. Many variations of the traditional narrowboat were built to coincide with local requirements. The most outlandish, and included here because of their shape, were vessels operating between London and Braunston on the Grand Union, which had locks 14 feet wide. These vessels were 'special wide narrowboats' but were built on narrowboat lines. They had barrel-shaped sides, but the bottom of the hull was 7 feet wide, which enabled them, with their narrow underwater cross section, to work on the Grand Union Canal with its basically narrowboat channels.

Certain fly-boats (for express work) and pottery boats were the lightest of the narrowboats built for normal trading. They were 69 feet long with a beam of 6 feet 10 inches, and the hulls were barrel-shaped with a slight 'V' bottom. In appearance they had very fine lines, especially at the fore-end. They carried a maximum load of 22 tons.

In Yorkshire certain wide waterways were connected with canals on which narrowboats usually worked. One of these was the Rochdale Canal, which regularly had narrowboats working over from Manchester to Sowerby Bridge. There the narrowboats, like the Mersey flats, had to stop, being unable to

85 *Clematis*, an ex-FMC motorboat, built in 1935, was photographed in 1964 when the vessel was operated by British Waterways. Although the exterior of the cabin is painted in a blue and yellow colour scheme without decoration, the brass rings and safety chain on the chimney along with the extension piece and pin of the tiller show pride by the skipper.

Nº10
20379
11613
FELLOWS, M
& CLAYTON, LTD
DENMARK

86 At Watford, a pair of boats are working up a lock of the Grand Union. Many fixtures and fittings can be seen on the vessels captured by this action shot of the camera. Note the skipper closing the bottom gate with the aid of the motorboat propeller. The spare exhaust chimney resting on the motor's roof is to lift the fumes high out of the face of the steerer when the need arises.

87 A motor boat and butty, owned by A Harvey Taylor of Aylesbury, are passing Tring. On the butty, along the top planks over the cargo hold, are a number of running blocks through which the rope passed and enabled the steerer of the butty to lengthen the distance between the two vessels, especially on long pounds.

88 ***Leeds*****, a 'wide' narrowboat, passes through Ironbridge lock during 1916. These vessels were used on the Grand Union Canal between London and Braunston for many years. Built on narrowboat lines with barrel shaped sides, the bottom of the hull was 7ft wide. Being used, but unfortunately not very clear in the photograph, is a pulley block system used to reduce the effort required by the horse to haul the vessel out of the lock.**

proceed any further because of the short locks of the Calder & Hebble Navigation.

The Lancashire side of the Leeds & Liverpool Canal accommodated narrowboats as far as Wigan, but the Yorkshire side could not take them again because of the short locks on that section of the canal. The only narrow canal in Yorkshire, the Huddersfield Narrow Canal, naturally accommodated the traditional full length narrowboat, working between Yorkshire and the many narrow canals in the North West. Here again, because of the short locks on the adjoining waterways, the narrowboats could not proceed any further east than Huddersfield.

To work on both wide and narrow beamed canals, a number of narrowboats only 61 feet long were built to work between the narrow canal and certain industrial towns and cities in West Riding which were situated on the broad-beamed, short-locked waterways. Only about three dozen of these 'West Riding' narrowboats (61 feet long, 22 ton cargo capacity) were ever built. They were both bulk liquid carriers and general cargo vessels, so they served a useful purpose and avoided much transhipment of cargoes. The last of these vessels, *Florence*, was built at Foleshill, near Coventry, in 1911 for J & E Morton of Milnsbridge. The only narrowboats built in the area were the two built by Albert Wood at Sowerby Bridge. They were of conventional size but they had higher than normal sides, and worked entirely on the Rochdale and Ashton canals. The last one, called *Reggie*, was launched in 1907.

The Chesterfield Canal, with locks both wide and narrow, had its own narrowboats, some with a short mast and square sail. Many of them regularly

89 The largest fleet of day boats in existence for many years, until their disposal in the mid 1970s, was owned by the British Steel Corporation. The final fleet of vessels consisted of both simple wooden-constructed day boats, as in the forefront, and a number of former butties with the superstructure and deck fittings removed.

worked on the tidal River Trent. In the Birmingham district, because of the tremendous amount of short haul work, including working to various narrow factory arms, they had a considerable number of 'day' boats, known as 'joey' boats. Most of these vessels were without cabins; if one was fitted it measured 5 feet 6 inches long with a maximum headroom of 5 feet 3 inches. Simply constructed, they were without the curves at the bow and stern that were usually associated with narrowboats. The majority of them were double-ended, useful when moored in a narrow factory arm, when it was simply a case of transferring the rudder from one end to the other. They were made with planks 15 inches to 19 inches wide, up to 40 feet long, and with the 10 inch by 3 inch keelson running the full length of the boat.

Another type of day boat was the 'railway' boat built for the Midland and the London North Western railways. Of similar construction to ordinary day boats, they had somewhat less severe lines, but were without cabins. When tugs became popular, long strings of these day boats were hauled, closely coupled, with usually only the last one being steered by a helmsman. In recent years a number of these have continued to be used as rubbish boats. They are moored up alongside a factory, and rubbish is tipped in until the hold is full; they are then towed away by a diesel tug to be emptied, while their places are taken by empty boats.

90 Moored alongside the towpath is a day boat of the Birmingham Corporation salvage department, who owned a considerable number until the mid 1960s. The cabin is smaller than a butty cabin, merely a shelter from the weather.

In the Black Country, the Wolverhampton level provided 54 miles of lock-free waterway. Having fewer limiting bridge hole restrictions than other narrow canals on the BCN network, it really was an internal semi-wide waterway. To operate on this section of water, vessels were built on the day boat pattern, able to carry as much as 50 tons. These craft were longer than the traditional narrowboat, ranging from 77 to 88 feet in length and with a beam of 7 feet 10 inches. So big were the largest holds at 75 feet long, that a conventional narrowboat could have been slotted into the hold. These day boats were commonly known as 'Hampton' boats and were used very profitably, mostly to transport coal.

The Bridgewater Canal was a wide waterway on which traditional narrowboats always worked, and via the Trent & Mersey Canal, which was

91 A typical South Wales, simply constructed canal boat: the photograph was taken on 1 January 1944 when the officials of the Cardiff City Council took over the Glamorganshire canal and railway.

a connecting waterway, and the $1\frac{1}{2}$ miles of the Rochdale Canal, boats came to the Bridgewater Canal from the Ashton, Peak Forest, Stockport and Macclesfield Canals. At Runcorn, a very robust and massive type of narrowboat was constructed, some roughly planked and devoid of any beautiful curves. Many of these boats were oversize and unable to pass through any narrow locks. They were known as 'wooden headers' because they had wooden bollards on each side of the bow deck. The motorboat when introduced had a butty-type stern and rudder, with the propeller shaft extending through the stern post. A pair of boats were able to carry a maximum load in the region of 64 tons, usually coal, but loads of potatoes were not unheard of. (One excellent example, *Hazel*, in private ownership, and now converted with living accommodation, can still be seen.)

Another different size of narrowboat was the 'trench' boat, belonging to the Shropshire Union. Of traditional design, these were 70 feet long, but only 6 feet 2 inches wide. They were built to operate through the narrow tub boat locks on the Shrewsbury Canal. Towards the end of their days a number were used for maintenance work on other canals. The last one in commercial use was the *Colonel*, named after a certain Colonel Padget, which could carry 18 tons.

The Welsh Canals had their own narrowboats, especially in the industrial area of the Swansea and Taff Vale Valleys, and not interconnected with the rest of the inland waterways. The narrowboats were unlike those that operated on the English canals. They were day boats, and mostly double-ended to avoid winding, but some of them had a small cabin provided for the crew to prepare meals. Their bows were without much shear, being almost vertical and rather ugly looking. They varied in size depending upon the size of the locks of the waterway they worked on, ranging from 60 to 65 feet in length, and they were much wider than normal narrowboats. For example the Glamorgan Canal boats were 60 feet long with a beam of 8 feet 6 inches maximum.

At the turn of the century, narrowboats working on the canals were an everyday scene, and it was assumed that this would continue forever. The problems of trading in competition with the railways had resolved themselves, and the canal traffic had settled down with steady trading figures. But in 1921 the Shropshire Union Canal Company's carrying concern was closed down, and this action came as a blow to narrowboat trading in general; it was in fact the start of a difficult trading period, for during the next decade came the country's first national strike, and its adverse effects were far-reaching. From then on the number of operators and narrowboats gradually declined, for the increase in the use of road vehicles was beginning to make itself felt.

92 *Chertsey*, a large Woolwich, town class motor boat, was built in 1937 for the then expanding narrowboat fleet of the Grand Union Canal. The vessel, of composite construction (metal with a wooden bottom), was inherited by British Waterways. They eventually disposed of her to a private buyer. The traditional cabin is longer than a butty to house the engine forward of the living section. Steering is by a 'Z'-shaped tiller, the rudder being practically concealed beneath the stern of the vessel.

93 This scene at Stoke Bruerne on the Grand Union, with many boats moored for the night, shows how busy the canal was for commercial transport. Nearest the camera is a loaded pair, and because of the nature of their cargo, the hold is not sheeted up. The cratches are not adorned with decorative paintings, but note the ropework, and since this is a late photograph of working boats, it shows them fitted with electric headlamps instead of paraffin for journeys through tunnels.

One boost to morale for the narrow canals was in 1929, when the Grand Union Canal Co Ltd came into existence, with the amalgamation of four independent canal companies. The new company immediately concentrated upon improving its network, and, with Government support, embarked upon a modernisation programme. In 1934 the Canal Company formed the Grand Union Canal Carrying Co Ltd, which it proceeded to expand, ordering a considerable number of new boats over the next few years. These new boats were built by various yards, including Harland & Wolff Ltd of Woolwich, James Pollock Sons & Co Ltd of Faversham, W H Walker & Bros Ltd of Rickmansworth, and W J Yarwood & Sons Ltd of Northwich.

In 1936 the company proceeded with its most ambitious plans, ordering 86 new pairs of narrowboats, the last of which were delivered late in 1938. The result was that a number of substantial classes of narrowboats were formed, notably the 'Star', 'Town' and 'Royalty' classes. This was the final big effort by a private company to improve trade by narrowboats. Although it is true that many more boats were to be built before narrowboat operations finally came to an end, never again were such large scale capital investments to be put into the narrowboats themselves.

Despite the GUCC Co's attempt to encourage transport by narrowboats, they were not wholly successful. They found it difficult to obtain enough satisfactory crews, so had on occasions to lay up large numbers of vessels. Throughout the 1939–45 War, the narrowboats did yeoman service, assisting

the war effort by providing an efficient transport service. Many of the boats had young female crews, who came from many different walks of life to work the narrowboats for the duration of hostilities.

On 1 January 1948, the inland waterways of Britain, with some exceptions, passed to the British Transport Commission, and became the direct responsibility of the Docks and Inland Waterways Executive. The remaining private carriers continued to trade with their vessels, while the fleet of vessels of the GUCC Co Ltd formed the basis of the nationalised fleet. On 25 October of the following year, the country's largest remaining private fleet of 135 vessels, belonging to Fellows, Morton & Clayton Ltd, was brought into carrying service with the Docks & Inland Waterways Executive. In the next decade competition from road transport became fierce, and narrowboats were generally regarded as an outmoded method of transport. Over the years the number of private operators decreased alarmingly, with many companies going out of business. One such was Samuel Barlow, a large operator who had specialised in the transport of coal.

Meanwhile the nationalised body continued to operate its fleet of narrowboats, obtaining new vessels to replace worn-out ones. In all it obtained about 40 new vessels, including the 'River' class boats, which were obtained with fibreglass hold covers, and were built in 1959 and 1960. This class of boat did not have the fine lines of the traditional vessel; it was of welded construction, and was built with vertical stems and underwater swim-bows. Another class of vessel that they had developed for general cargo work was the 'Admiral' class, built by both Pimblott and Yarwood of Northwich in 1960 and 1961.

94 The British Waterways narrowboat fleet, consisting of several hundred vessels from two former carriers, were worked hard for a number of years after nationalisation, as this busy scene at their Tyseley Depot in Birmingham during 1957 shows. Their problems were many, with continually increasing competition, but hopes ran high with the delivery of new vessels. Unfortunately even the weather was against them, for the hard winter that froze the canals solid for many weeks in 1963 finally killed off narrowboat operations for the nationalised body.

95 This pair of river class vessels was obtained by British Waterways in 1960, of welded construction and without the fine lines of the traditional narrowboat. The majority of them have now found a new lease of life being used as camping boats during the summer months for youth organisations; the modern fibreglass hold covers are very useful for this new job.

Yarwood actually built the last two pairs, which were *Mountbatten* and *Keppel*, and *Lindsay* and *Jellicoe*.

Narrowboat operations continued, but with ever increasing difficulties, and the fleet of the British Waterways Board, an independent body of the former nationalised group formed in 1962, was gradually run down until it finally ceased activities altogether. Not all the craft were sold or used on maintenance duties, for on 14 September 1963 the Board leased 25 pairs to Willow Wren Canal Transport Services Ltd, who continued to operate until 1970. More and more private operators, by now very few in number, went out of business. Some of them were very old established carriers like Thomas Clayton (Oldbury) Ltd, who, after nearly a century of trading, ceased their canal activities in 1966. They had specialised in the transport of bulk liquids in wooden 'tank' boats.

96 In 1964 at Wolverhampton Top Lock is *Umea*, a tank craft of Thomas Clayton (Oldbury) Ltd who specialised in transporting bulk liquids. *Umea*, built in 1939 of wooden construction, was typical of their fleet; metal vessels were tried but not liked by the crews. The company ceased trading with narrowboats in 1966, but a number of their vessels, both motor and butties, have been preserved.

In the 1960s a company financed by enthusiasts was formed, and after a while it was obliged to diversify to passenger transport in order to survive. In the Midlands the last regular working of the old pattern was the transport of phosphorous effluent, from Oldbury to Albion Junction by Alfred Matty & Sons Ltd. This ended in 1975. A number of narrowboats are now employed in the summer months as camping boats for youth organisations, while during the winter months some of these vessels are employed on domestic coal trade, travelling along the canal system selling the cargo. The last regular traffic still in operation, using traditional vessels, is for the transport of lime juice from

97 An excellent example of a narrowboat kept in the best of condition, and working: ex-FMC's *Jaguar* is loaded with coal on the Ashby Canal.

JAGUAR
JAGUAR

98 Of welded steel, and functional lines, this is one example of a modern narrowboat, used for transporting crated pottery daily between two works of a company located alongside the Caldon Canal.

Brentford to Boxmoor, using one pair of boats usually on twice weekly runs. The only other traffic employing a few vessels is the movement of concrete aggregates several miles on the River Soar section of the Grand Union Canal.

Trying to solve the problem of transporting pottery without breakages resulted in a new approach to narrowboat operations by a canalside company. In 1967, Johnson Bros (Hanley) Ltd launched a specially designed vessel, *Milton Maid*, to transport pottery in wire crates between two factories 4 miles apart. The crates were wheeled directly onto the flat deck of the boat at one factory and wheeled off at the other. The operation proved satisfactory, and a second vessel, *Milton Queen*, was brought into service in 1973. The vessels are made of steel and work as day boats, so they have no cabins. They are powered by petrol engines, carry 20 tons of pottery each, and are in continual use. This is a good example of a modern adaptation of an old system. In October 1978 the company extended their waterway operations by launching a third vessel, *Milton Princess*, which was 70 feet long and powered by a 24hp Russel Newberry diesel engine.

Maintenance Vessels

MOST river navigations and canal cuts are over two hundred years old. They are in a usable condition today only because regular maintenance work has been undertaken over the years. To keep a waterway in a navigable state, it requires constant attention, for water continues to wear things away. Even still water will rot many types of material over a period of time. Even so, a great number of lock chambers in existence now were constructed when the waterways were opened, although the gates, made of best English oak, will have been changed at least every fifty years, and in many instances more often than that. Really it is only because of constant use and regular maintenance that the inland waterways have been kept alive, for left undisturbed they deteriorate very quickly indeed. With choking weed impeding the passage of vessels, the channel silts up, so that, in only a short time, it reverts to a natural state.

Any waterway company, whether responsible for a navigation or a canal, has always had to undertake maintenance work and provide facilities and equipment for the users. The pattern of organisation and procedure that evolved over the years has continued virtually unchanged to the present day. Throughout the system a chain of maintenance yards exist. Most of the yards are small and deal with everyday routine matters. Interspaced with these are larger depots, with extensive workshops and repair facilities. Some of the larger yards also build and repair maintenance boats, and the central repair depots make lock gates and moving bridges.

The basic function of the maintenance section is to provide a good navigable channel for the passage of all vessels, ensuring that as little delay as possible is experienced when major work is undertaken. They have to prevent the erosion of banks and remove the silt, mud and debris that accumulates in the water channel. Repairs and renewals to aqueducts, bridges, locks, tunnels, sluices and weirs, etc, are all part of the work that has to be undertaken by the maintenance section. This requires the skilled services of bricklayers, carpenters, joiners, labourers and masons, along with workers from other allied trades, ranging from sail and block makers in the past to engineers and electricians now, as well as boatmen to move the vessels that are used.

Naturally there has always been a need for vessels for maintenance work, and these initially just satisfied the most basic requirements. Usually they were the flat-bottomed, shallow box type, able to support and transport a few tools a short distance along the waterway. From such humble beginnings, the vessels of the maintenance department have progressed through periods of wood,

99 A wooden-constructed general purpose maintenance boat of the Leeds & Liverpool Canal, in the 1930s.

iron, steel, and steam power, to hydraulics and lightweight metals. Some of the equipment now in use is most sophisticated, and incorporates the latest developments in engineering.

Because of the continual need to provide a good clear channel, there has, from the very beginning, been a constant need for dredging operations. The earliest dredger was nothing more than a strong wooden, punt-shaped vessel on which was fitted a small manually operated crane. To this was attached a spoon-shaped scoop, which was lowered into the water and, with the aid of physical labour, spooned the mud from the bottom of the canal. The mud was either tipped into a vessel alongside or piled on the canal bank. Surprisingly on certain waterways, this old system of spoon dredging continued until quite recently, in particular on the Birmingham Canal Navigations. Eventually steam power was harnessed to the spoon dredgers, and equipment became more sophisticated. Mobile cranes with bucket grabs made the work easier, while larger units that used an endless chain system with scoops attached was the ultimate.

Some of the dredgers, especially those used on rivers, were of large proportions, discharging the mud from the bottom of the river down a chute into waiting mudhopper vessels. These mudhoppers in turn, after moving to a dispersal area, were discharged by mechanical grabs. Now on certain waterways large mobile pump units are provided to pump sludge out of the mudhoppers. In recent years diesel-powered hydraulic equipment has increasingly been brought into use, but a considerable number of old dredgers are still to be seen, although now they are diesel-powered instead of having the original steam engines.

All types and sizes of vessels have been provided to keep the waterways open. Flats or pontoon-type craft transport lock gates and lifting equipment, whilst other vessels are employed to carry men, materials and machines. Vessels with platforms enable painters to paint the undersides of metal bridges, and there are boats for specialists such as divers and joiners. More and more craft of the rectangular compartment-boat shape are being built, and these are pushed along the waterway by small powerful 'push-pull' tug units.

One of the most serious hazards to navigation in the past was ice. A frozen waterway stopped traffic immediately, and the revenue from tolls was lost. The directors of waterways, intent on keeping money coming in, soon introduced ice-breakers. The ice-breakers, used for many years on most canals, were simply-designed boats with reinforced bows and metal-sheaved hulls. The boats varied in size, but usually were about 7 feet wide and 30 to 35 feet long, and very shallow draughted. They carried a platform with at least two

100 Spoon dredging by the old method, using plenty of muscle power, on the Leeds & Liverpool Canal at Clayton-le-Moor.

101 The early steam bucket dredger, *Widnes*, built in 1881, below Bewsey lock on the St Helen's Canal in July 1957. The dredging unit was manufactured by Priestman of Hull. The steam engine also drives the vessel's propeller by means of bevel gearing and a shaft.

102 The modern method of transporting dredgings on narrow canals is with the use of specially built mud hoppers that are pushed along by small but powerful diesel tugs. The hydraulically operated land-based grab quickly empties the hoppers.

103 A former steam powered large inland waterway dredger: now powered by a diesel engine, it is similar to many used by the British Waterways Board on river navigations.

104 A steam dredger at work on the Selby Canal early this century: the spoil dredged up is being loaded into the 61ft 'West Riding' narrowboat alongside.

105 When the canal froze, traffic stopped and so did the revenue, and in an attempt to keep traffic moving ice-breakers were introduced. The photograph shows one underway, being rocked from side to side hauled by ten horses. Of wooden construction, and sheaved on the exterior by metal plates, they did a useful job for many years.

uprights, connected by a horizontal bar. A number of men would stand on the platform, gripping the bar, and rock the boat. Meanwhile a number of horses attached to the boat by a line would proceed along the tow path as fast as possible. The bows of the boat, bouncing up onto the ice, would break it, and so the boat proceeded along the waterway, bouncing up and down and lurching from side to side. Having broken a channel through the ice, the ice-breaker would be followed by a number of cargo boats, who may have been held up for a number of days and were anxious to complete their journey.

On many waterways when the use of iron-hulled steam tugs became common, these were used to smash their way through the ice to provide a passage for the cargo vessels. Now, only a few waterways are regularly used during the winter, and these commercial waterways rarely now experience a stoppage due to ice, one reason being the extensive use of canal water abstracted by industry for cooling purposes. When this water is returned to the waterway it is usually several degrees above the temperature of the water in the channel, and thus prevents the canal from freezing over.

Passenger and Pleasure Boats

PASSENGER BOATS were provided on numerous inland waterways both by canal companies and by independent operators. In both cases their operation provided additional revenue for the canal companies. The passenger boats were in operation from the earliest days, providing a regular service between canal and riverside towns and villages, the only alternative being uncomfortable journeys in stage coaches, bouncing over unmade roads. Travel by water was smooth, pleasant and in many cases quicker.

On many canals, especially those with few locks, an express service operated, with the packet boat drawn along by teams of cantering horses, ridden by postilions, resplendent in company uniform. The horses were changed at frequent intervals to ensure the speedy movement of the boat. The packet boat in most cases enjoyed top priority passage, as it had the right of way on all occasions, including at locks and bridgeholes. In fact so determined was the company to show the priority of movement the passenger boats

106 This photograph shows the old packet boat, 76ft long by 5ft 11in beam, that operated on the Lancaster Canal, carrying passengers between Kendal and Preston. In operation its average speed was 10mph, with a change of horses every 4 miles. *Waterwitch II*, built in 1839, after the withdrawal of passenger services was used for many years as an inspection boat before being finally broken up.

107 The canals of Scotland continued to be used for passenger transport after the general demise of the packet boat. Many steamers were introduced for passenger carrying and continued to operate most satisfactorily for many years. The fact that the Forth & Clyde Canal had no fixed bridges over its entire length enabled vessels with high superstructures to operate. *Gipsy Queen*, 68ft long and 19ft wide, was one of the famous Queens that operated during the last century.

enjoyed on the Bridgewater Canal, that they had a forbidding looking 'S' shaped blade fitted at the fore-end, which would cut through the tow rope of any other craft that dared to get in the way and impede their passage.

The packet boats were usually lightly built, with fine lines and pointed bows to assist their speedy movement through the water. The passengers were accommodated as befitted their station in life, with first, second and, on occasions, even third class saloons. Standards of comfort varied from the high quality of first class travel, with curtains, upholstered seats and carpeted floors, to wooden benches for the cheaper class of travel. Facilities also varied, some providing meals and refreshments at all times of the day, while others stopped at canalside inns and eating houses. In the 1770s, first-class travel on the horse-drawn packet boats usually cost about one penny per mile, while the average speed maintained throughout the journey was between 3 and 4 miles

108 During the Victorian period pleasure cruises on the rivers became a popular pastime and many steam craft were built for the trade. This photograph, taken in 1904, shows the steamer *Empress* on the River Trent at Nottingham. The steel-hulled vessel, built at Gainsborough in 1896, was 80ft long and 14ft wide, with seating capacity for 242 passengers.

per hour, depending on the number of locks on the route. For nearly a hundred years the horse-drawn packets continued to operate, although, after the coming of the railways, many provided more of a pleasure trip than a transport service.

In the 1830s the passenger service on the Forth & Clyde Canal began to be undertaken by steam packets, at first by paddle steamers, and then by screw-driven vessels. Afterwards steam packets were used on other canals, replacing the horse-drawn vessels. But as time passed, fewer passenger services remained on the canals. The Scottish canals, renowned for their views of natural beauty, remained the chief operating area for steam packets. One well known company, David Macbrayn, with a fleet of vessels playing on many routes on the west coast of Scotland, had *Linnet*, 86 feet long, specially built for operation on the Crinan Canal in 1886. The last passenger steamer on the

109 On the Staffs & Worcs Canal in 1910, the directors' steam launch is being used for a workpeople's annual outing.

canals in Scotland did not stop work until 1928, while in England one isolated service continued until 1932, operating on the Gloucester & Berkeley Canal.

Passengers were transported on rivers by steam powered vessels well before their general use on canals. In fact passenger services on some river navigations became a very important means of transport for some considerable time. In 1815 on the River Ouse, coach services from West Yorkshire connected at Selby with river steamers which operated from there to Hull. The following year, on 25 April 1818, the passenger boat *Waterloo*, a wooden steam packet, inaugurated steam passenger services between York and Hull. In 1822 there were at least five steam packets operating on the River Ouse, and these services continued until well after the spread of railways in the area.

During the mid-nineteenth century the packet boats on the canals were gradually withdrawn, until really only those providing pleasure trips remained, but many steamers on rivers went on operating most satisfactorily, especially if the river ran through beautiful countryside. On the River Thames, Salter's steamers maintained a twice daily service between London and Oxford. This had started in the 1830s, and their last steamer, the *Clivedon*, was built in 1931.

While the passenger services of necessity came to an end, trips for pleasure increased as they provided a social service. At the turn of the century, for the first time, the lower classes had a rather increased spending power and a desire to travel. Many steamers were built for the trade on such rivers as the Dee, Severn, Thames, Trent and Warwickshire Avon. The majority of these steamers had a distinctive hull shape, with a well raked fore-end and a round counter stern, the same basic lines, indeed, as the pleasure yachts of that period owned by the well-to-do.

Throughout the Victorian and Edwardian period, it became usual for canal companies to have their own passenger craft, used solely for inspection trips by

110 Generally on the inland waterways, pleasure cruising was slow to develop. A lack of demand resulted in a shortage of vessels available, and so whenever a cruise was undertaken, it usually had to involve the use of a commercial carrying vessel. A good example of this practice is captured here at Brendon Staunch in the late 1890s, where a pleasure party have hired the lighter *William Murrell*.

their engineers and the directors. These vessels, specially built, were initially horse drawn and later steam powered. They were of the most elegant designs and well built, using all the expertise of the craftsman boatbuilder, carpenter and upholsterer. The directors' inspection boats certainly were built to the highest possible level of luxury, comparable with the best of railway carriage saloons, even those built for royalty. The interiors had monogrammed carpets and etched plate glass windows, with leather upholstered seating and mahogany panelling. Some of the most notable examples were the Weaver Navigation's *Water Witch*, the Worcester & Birmingham Canal's *Swallow*, and the Staffordshire & Worcester Canal's *Lady Hatherton*. All three vessels have now gone, and their like will never be seen again.

On many rivers, small boats, such as punts and rowing boats, have been used ever since the Victorian period, whilst the steamers became a much acclaimed way of spending a pleasant afternoon, and introduced many people to inland waterways. Sunday Schools and other charitable organisations,

111 Because of a surplus, many hundreds of narrowboats were sunk during the 1950s. Then because of the wind of change resulting from a demand for narrowboats for pleasure cruising purposes, many were salvaged during the 1960s. Here can be seen one method of lifting narrowboats from the bottom of the canal. These craft, once a common sight on the Leigh branch of the Leeds & Liverpool Canal, were the ones that transported the coal containers which were 6ft by 4ft 6in.

112 After the last war when canal cruising started to become an accepted way to spend one's time, on many waterways it became the practice to use either converted life-boats or vessels whose design was based on the Broads and Thames cruisers. Here on the Huddersfield Broad Canal is a canal cruiser with very similar lines to the river cruisers.

113 On the Calder & Hebble Navigation in the early 1960s is the British Waterways vessel *Fair Maiden*. During the 1950s a demand by the public for day cruising facilities resulted in many former working boats being converted for passenger carrying. *Fair Maiden*, converted at Goole, was a former Leeds & Liverpool Canal motor 'short boat'.

114 The ultimate for a canal enthusiast is to own a genuine converted narrowboat for cruising, with the traditional cabin incorporated. Alternatively, there is the cruiser specially built with the lines and characteristics of the genuine article – with round counter, Z-shaped tiller, and aft cabin doors to a cratch at the fore-end. Here is a good example of the best that money can buy to achieve the ultimate; the lines of the superstructure are good, housing no doubt all the modern luxuries available for life afloat.

115 This pleasure cruising vessel was built on a steel hull for strength, and is very functional, but is not a copy of the fine lines of the working boats. It was designed to give every luxury, for incorporated in the fittings are all the modern refinements. Powered by a reliable diesel engine, it is able to cruise throughout the network of inland waterways.

116 In recent years passenger carrying vessels by the score have been introduced, some former working narrowboats converted, and others like this vessel on the Sheffield & South Yorkshire especially built on the tried and tested narrowboat lines.

LORD OF THE RINGS
beaver

117 This steel, diesel powered narrowboat cruiser, is perhaps not as traditionally lined as some, but is very popular, incorporating good accommodation and handling abilities.

intent on providing an unusual but enjoyable afternoon out, arranged 'A second best' – a boat trip along a canal. As no passenger boats were available for such canal trips, the organisers would hire a working boat and provide it with bunting, bench seats and perhaps a piano. So they enabled many to sample the delights of canal cruising. These trips were a success, and the canals came to be known as places where, despite their image of being work-a-day places, one could find enjoyment with a touch of adventure.

Increasingly, privately owned craft were seen on the rivers, most of them owned by wealthy people, and based on the Victorian elegance of mahogany and teak. A few owners were adventurous and strayed onto the canals, but they were the exception rather than the rule, for canals were considered the 'place where barges worked'. From the early 1890s the Broads had become popular for sailing vessels, and all the time more and more people took to the water. In the 1930s when the first canal hire craft were made available, hire craft were already established on the River Thames and the Broads.

After the war, with more money being available, along with more free time for practically everyone, it was natural to expect an increase in the use of rivers and inland waterways for pleasure. The demise of the working narrowboat in the midlands, and an appreciation of the environment certainly helped to accelerate the widespread use of the canals for pleasure cruising. Hire boats in use in the 1950s were few in number, but by the '60s their numbers had increased considerably, while in the '70s they have spread so quickly that restrictions on their numbers have been imposed in certain districts.

Many of the first hirers of canal boats were so delighted with the experience that they went ahead and bought their own pleasure craft. Because so many people enjoyed 'messing about in boats', their numbers grew so much that in recent years dozens of marinas have been opened to cater for this increase, which only a few years ago would have not been thought possible. Pleasure craft on the canals range from converted narrow working boats and fibreglass cruisers to specially built steel-hulled cruisers that move extensively on the several thousand miles of inland waterways open to them.

118 The cost of wooden boats has increased considerably in recent years, and so the fibreglass cruiser has become very popular for canal cruising. It is low on maintenance, and more reasonably priced than its steel counterpart, but offers the same advantages. A vessel like *Beaver*, fitted out to a high standard, enables four people to cruise in comfort throughout the network of inland waterways. The inboard diesel engine ensures reliability, power when required, and, to an extent, economy. The 'Z' drive gives excellent handling properties.

Directory of Historical Canal and Inland Waterway Craft

The following is a list of historical vessels of particular interest, restored and otherwise, in existence, owned by museums, restoration groups etc. A considerable number of historical vessels, especially narrowboats are owned by individuals; these are not included.

Type of vessel	Name	Length	Details	Owner or base where vessel can usually be viewed
Barge	*Bigmere*	71 feet	Ex Bridgewater Canal	The Boat Museum, Ellesmere Port.
Barge	*Black Prince*	45 feet	Fenland of carvel construction	Museum of Technology, Cheddars Lane, Cambridge.
Barge	*JJRP*	37 feet	Ex Taw & Torridge estuary	North Devon Maritime Museum, Odun Road, Appledore, Devon.
Barge	*Constable*	47 feet	Ex River Stour	River Stour Trust, 30 Normandy Way, Bures, Suffolk.
Flat	*Mossdale*	70 feet 4 inches	Ex Mersey & Weaver	The Boat Museum, Ellesmere Port.
Icebreaker	*Sarah Lansdale*		Ex Manchester, Bolton & Bury Canal.	The Boat Museum, Ellesmere Port.
Icebreaker	*Marbury*		Shropshire Union	Shropshire Union Canal Society, Market Drayton.

Type of vessel	Name	Length	Details	Owner or base where vessel can usually be viewed
Icebreaker	*Middlewich*			Ironbridge Gorge Museum, Telford, Shropshire.
Keel	*Ethel*	57 feet 6 inches	'West country' size wooden/motor	The Boat Museum, Ellesmere Port.
Keel	*Comrade*	61 feet 6 inches	Sheffield size steel/sailing	Humber Keel & Sloop Preservation Society, Beverley.
Keel	*Annie Maud*	61 feet 6 inches	Sheffield size wooden/sailing	River Ouse, York.
Leeds & Liverpool boat	*Scorpio*		Wooden construction	The Boat Museum, Ellesmere Port.
Leeds & Liverpool boat	*Pluto*	61 feet	Wooden construction	The Boat Museum, Ellesmere Port.
Leeds & Liverpool boat	*George*		Wooden construction	The Boat Museum, Ellesmere Port.
Maintenance vessel	*Jaws*		Weed cutter, ex Rochdale Canal	The Boat Museum, Ellesmere Port.
Maintenance vessel	*Bertha*		Steam dredger built in 1844	Exeter Maritime Museum of Boats.
Narrowboat	*Chiltern*		FMC wooden/motor boat	The Boat Museum, Ellesmere Port.
Narrowboat	*Friendship*		No. One Horse boat	The Boat Museum, Ellesmere Port.
Narrowboat	*Gifford*	71 feet 6 inches	Thos Clayton Horse tar boat	The Boat Museum, Ellesmere Port.
Narrowboat	*Puppis*		Iron butty, Ex Grand Union	The Boat Museum, Ellesmere Port.

Type of vessel	Name	Length	Details	Owner or base where vessel can usually be viewed
Narrowboat	*Pheobe*	70 feet	Iron/BCN joey	The Boat Museum, Ellesmere Port.
Narrowboat	*Nos 22 & 39*		Wooden day boats ex BCN	The Boat Museum, Ellesmere Port.
Narrowboat			Starvationer, ex Worsley	The Boat Museum, Ellesmere Port.
Narrowboat	*Viceroy*		FMC steamer	On charter.
Narrowboat	*Northwich*		FMC composite butty	Waterways Museum, Stoke Bruerne, Towcester.
Narrowboat			Starvationer	Lound Hall, Mining Museum, Nottingham-shire.
Narrowboat	*Vienna*		FMC composite butty	Cheddleton Flint Mill Museum, near Leek, Salop.
Narrowboat	*Bessie*		Iron day boat	The Black Country Museum, Tipton Road, Dudley.
Narrowboat	*Admiral Beatty*		Wooden day boat	The Black Country Museum, Tipton Road, Dudley.
Narrowboat	*Birchills*		Wooden day boat	The Black Country Museum, Tipton Road, Dudley.
Narrowboat	*Ivy*	58 feet 6 inches	Neath Canal boat	Welsh Industrial & Maritime Museum, Cardiff.
Passenger boat *Doreen*		30 feet	ex-Windermere pleasure motor boat	Calder Navigation Society, Calder & Hebble navigation.

Type of vessel	Name	Length	Details	Owner or base where vessel can usually be viewed
Passenger boat	*Hero*	35 feet	Ex Thames steamboat	John Players Ltd, Nottingham.
Passenger boat	*Lady Betty*		Steam launch	Exeter Maritime Museum of Boats.
Passenger boat		70 feet	Broads pleasure steamer	Veteran Steamship Society, 8, High Street, Dunnow, Essex.
Passenger boat	*Brinksome*	50 feet	Selection of vessels showing steamboat development. Vessels built between 1850 and 1912	Windermere Steamboat Museum, Rayrigg Road, Windermere.
Passenger boat	*Esperence*	65 feet		
Passenger boat	*Raven*	71 feet		
Passenger boat	*Bart*	27 feet		
Passenger boat	*Lady Elizabeth*	18 feet		
Passenger boat	*Otto*	53 feet		
Passenger boat	*Swallow*	45 feet 6 inches		
Passenger boat	*Canfly*	26 feet		
Sloop	*Amy Howson*	61 feet 6 inches	'Sheffield' size Yorkshire sloop, steel/sailing	Humber Keel & Sloop Preservation Society, South Ferriby.
Tug	*Worcester*	44 feet 6 inches	Ex Worcester & Birmingham Canal	The Boat Museum, Ellesmere Port.
Tug	*Reliant*	100 feet	Steam tug ex Manchester Ship Canal	National Maritime Museum.
Tub boat			Leading tub boat, ex Bude Canal	Historical & Folk Exhibition, Bude, Cornwall.
Tub boat		20 feet	Tub boat fitted with wheels, ex Bude Canal	Exeter Maritime Museum of Boats.
Tub boat		20 feet	Wrought iron tub boat, ex Shropshire Tub Boat Canal	Ironbridge Gorge Museum, Telford, Salop.

Type of vessel	Name	Length	Details	Owner or base where vessel can usually be viewed
Trow	*Spry*	73 feet	Scuttled	Worcester.
Tug	*St Canute*		Steam tug built in 1931	Exeter Maritime Museum of Boats.
Thames sailing barges	*Thalotta* *Sir Alan Herbert*		Built 1906 Built 1926	East Coast Sail Trust, Maldon, Essex.
Thames sailing barges	*Nellie Parker* *Asphodel* *Conway*			Dolphin Sailing Barge Museum, Crown Quay Lane, Sittingbourne, Kent.
Thames sailing barge	*Cambria*		Trading under sail until 1971	Maritime Trust, The Esplanade, Rochester, Kent.
Thames sailing barges	*Pudge* *Centaur*	82 feet 82 feet	Built 1922 Built 1885	Thames Sailing Barge Club.
Wherry	*Albion*	60 feet	Norfolk wherry built 1898	Norfolk Wherry Trust, Norwich Road, Lingwood, Norfolk.

Index

The figures in italics refer to illustration numbers; all other figures denote page numbers.